PRAISE FOR *Fruitful Labor*

"*Fruitful Labor* is a delightful book, full of practical advice and deep thinking about ecology and true sustainability. I highly recommend it for anyone interested in food and farming, but especially for young farmers looking to build their skills while gaining wisdom from someone experienced and respected in the field."

—BEN HARTMAN, author of *The Lean Farm* and *The Lean Farm Guide to Growing Vegetables*

"Mike Madison offers new and aspiring farmers a book outside the usual vein of small farm narratives and how-to tomes. Discontented with both formulaic prescriptions for the idealized family farm and mega-data studies that sacrifice particularities for trends, Madison instead digs deep into the three decades of farming history on the California plot where he and his wife raise more than 200 plant varieties, ranging from vegetables to flowers to olives—and no sacred cows. Madison puts nothing other than nature itself on a pedestal, and he questions his every decision by way of an ecological mirror that reflects back on him without embellishment or distortion.

"He confesses that he is not enamored with the current celebrations of mission statements, goals, and strategies. Rather, he describes the evolution of his family farm as a timeline without a road map—decision points on a long chronology, all informed by unhurried observation. His story is one of searching out hard-won possibilities through perseverance more than strategy.

"New farmers would be wise to take a day and travel with Madison through the course of his thirty-plus years, learning what lenses to use in examining each ecological, economic, and community-minded decision that all farmers face."

—PHILIP ACKERMAN-LEIST, professor of sustainable agriculture and food systems, Green Mountain College, and author of *A Precautionary Tale*

"Mike Madison writes from a place of knowing that one acquires only through lived experience. The deep ecology he prescribes, which 'advocates the rights and values of all species regardless of their utility to human enterprises,' should be the central principle of food and farming systems. Akin to the creature in the crystal river in Richard Bach's book *Illusions*, Mike stopped clinging a long time ago and let the current take him to a higher plane of thought and deed. Proof of this is sprinkled throughout *Fruitful Labor*.

"This book is a must-read for those embarking on their journey into farming and for all others who are remotely connected to food and farming, which is all of us."

—SRIDHARAN (SRI) SETHURATNAM,
director, California Farm Academy

Fruitful Labor

Other Books by Mike Madison

Walking the Flatlands:
The Rural Landscape of the Lower Sacramento Valley

Blithe Tomato

Fruitful Labor

The Ecology, Economy, and Practice of a Family Farm

MIKE MADISON

Chelsea Green Publishing
White River Junction, Vermont

Project Manager: Alexander Bullett
Project Editor: Benjamin Watson
Proofreader: Helen Walden
Designer: Melissa Jacobson

Printed in the United States of America.
First printing February, 2018.
10 9 8 7 6 5 4 3 2 1 18 19 20 21 22

green press INITIATIVE

Chelsea Green Publishing is committed to preserving ancient forests and natural resources. We elected to print this title on 100-percent postconsumer recycled paper, processed chlorine-free. As a result, for this printing, we have saved:

16 Trees (40' tall and 6-8" diameter)
22 Million BTUs of Total Energy
3,755 Pounds of Greenhouse Gases
15,403 Gallons of Wastewater
1,295 Pounds of Solid Waste

Chelsea Green Publishing made this paper choice because we and our printer, Thomson-Shore, Inc., are members of the Green Press Initiative, a nonprofit program dedicated to supporting authors, publishers, and suppliers in their efforts to reduce their use of fiber obtained from endangered forests. For more information, visit: www.greenpressinitiative.org.

Environmental impact estimates were made using the Environmental Defense Paper Calculator. For more information visit: www.papercalculator.org.

Our Commitment to Green Publishing

Chelsea Green sees publishing as a tool for cultural change and ecological stewardship. We strive to align our book manufacturing practices with our editorial mission and to reduce the impact of our business enterprise in the environment. We print our books and catalogs on chlorine-free recycled paper, using vegetable-based inks whenever possible. This book may cost slightly more because it was printed on paper that contains recycled fiber, and we hope you'll agree that it's worth it. Chelsea Green is a member of the Green Press Initiative (www.greenpressinitiative.org), a nonprofit coalition of publishers, manufacturers, and authors working to protect the world's endangered forests and conserve natural resources. *Fruitful Labor* was printed on paper supplied by Thomson-Shore that contains 100% postconsumer recycled fiber.

Library of Congress Cataloging-in-Publication Data
Names: Madison, Mike, 1947– author.
Title: Fruitful labor : the ecology, economy, and practice of a family farm / Mike Madison.
Description: White River Junction, Vermont : Chelsea Green Publishing, [2018]
Identifiers: LCCN 2017048973| ISBN 9781603587945 (hardcover) | ISBN 9781603587952 (ebook)
Subjects: LCSH: Farm life—California—Sacramento Valley. | Family farms—California—Sacramento Valley.
Classification: LCC S521.5.C2 M34 2018 | DDC 338.109794/53—dc23
LC record available at https://lccn.loc.gov/2017048973

Chelsea Green Publishing
85 North Main Street, Suite 120
White River Junction, VT 05001
(802) 295-6300
www.chelseagreen.com

Contents

Chapter 1

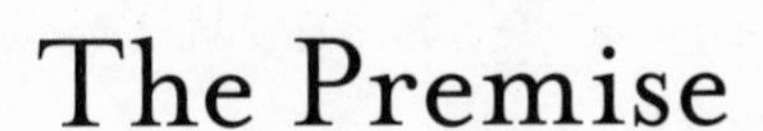

The Premise

For the last 4,000 years the commonest human occupation has been small-scale agriculture. Although it has been a few generations since that was the case in the United States, the image of the small family farm is still a powerful icon of our cultural identity. Urban dwellers weary of the chaos of city life, and tech workers in their cubicles spending their days in extremes of abstraction, dream of a simple, agrarian livelihood. It is not a thousand-acre industrial farm that they are thinking of, although such farms command the majority of farmed acreage in North America, but a small, diversified farm that operates on a comprehensible human scale. And each year thousands of people, mostly young, mostly inappropriately educated, start small farms with hope and courage. Most of these farms fail, some quite rapidly, but I imagine that regrets are few.

Given the ubiquity and long history of small farms, it is surprising how little has been published describing the operation of such a farm in a logical and thorough way. There are plenty of agricultural memoirs that favor a narrative—romantic, or lyrical, or amusing—of farm life, and there is much to be learned from these, but they are unsystematic and unquantified. And opposed to these is a large academic literature that is based mostly on statistical analysis of aggregated data in which the individuality of a particular farm is entirely lost.

My intent in this work is to describe the operation of a successful small farm over a period of thirty years. All agriculture is local, and the particular details of my operation might not be applicable elsewhere, but the basic variables are universal, and every farmer has to solve the same set of problems in whatever way works in his or her circumstances.

I should point out that my approach to farming is a contrary one, and my ideas of a good way to farm are at odds with mainstream farmers. There are other farmers with philosophies similar to mine, but all of us are operating at a small scale, and our collective acreage is minuscule in the big picture of American farming. So be advised that what I am presenting here is not the orthodox story.

The Scope of Agroecology

Each year, starting in mid-April, I plant 200-foot rows of cucumbers at ten-day intervals for 70 or 80 days. As the cucumbers come into readiness, I harvest them and bring them to the local farmers market on Saturday mornings and Wednesday afternoons, where I offer them for sale at three for a dollar, or, if business is slow, four for a dollar. I may barter cucumbers with some of the other farmers for produce that I don't grow myself, such as strawberries or avocados. Cucumbers that are damaged I bring home to feed to my chickens, and the rest of the unsold cucumbers I donate to a soup kitchen that feeds homeless people. This enterprise is so simple that any Chinese peasant or medieval European serf would readily understand it without further explanation.

And yet, the enterprise is crowded with questions. Would I be better off growing an expensive hybrid variety of cucumber rather than the old open-pollinated strain that I grow? Would the cucumbers benefit from trellising with respect to increased yield or freedom from damage? Would this pay? Could I use a sturdy crop—corn or sorghum—for the trellis instead of wood or bamboo poles? Would this require more water? Instead of plowing down my cover crop of beans and oats and working up the whole field, would it be advantageous to just mow the cover crop and till only narrow strips for the cucumbers? Could I do this with equipment that I already have, or would I have to purchase or fabricate strip-tillage equipment? How would strip tillage affect the populations of burrowing rodents—gophers, voles, and squirrels—that damage the cucumbers? Would the straw from the mown cover crop benefit the cucumbers by keeping them off the soil, or would it harbor insect pests? Should I add a band of compost, or gypsum, to the tilled strips to improve soil tilth and fertility? If I mow the cover crop, would I have to harrow the ground to plant a cover crop the following year, or could I just sow the seeds into the straw from the previous year? Would it be more profitable to grow a more valuable crop—for example, Charentais melons—instead of the cucumbers? Can I grow the cucumbers in the same field several years in a row, or should I rotate them with an unrelated crop, such as sunflowers or tomatoes?

These questions, and hundreds of others like them, find a home in the formal field of agroecology. The term "agroecology" is used in two senses. In the narrow sense, it refers to the application of the concerns and techniques of traditional wild-lands ecology in an agricultural setting. The investigator is interested in nutrient cycling, water relations, energy flow, canopy structure, interactions among species, population dynamics, phenology, and succession. Much of this literature may be found under the heading of "crop ecology."

In addition to these classical ecological subjects, agroecology in the broad sense expands to encompass inquiries into the economic and social context of the farm: for example the regulatory environment, the structures of markets, the operation of subsidies and tariffs, the role of publicly funded irrigation projects, political lobbying efforts of the farmers, access to credit, cooperation and competition, the flow of information, the influence of land-grant universities, and the operation of corporate cartels.

Whether a particular investigation is framed in the narrow sense (crop ecology) or the broad sense of agroecology is especially significant in two respects. The first has to do with scale. A purely crop-ecological study of a set of practices with respect to irrigation, soil management, fertility, and pest control might find them to be scale-invariant, and thus equally valid for a farm of twenty acres and a farm of two thousand acres. But adding the social and economic dimensions to the analysis would show it to be sensitive to scale, and the same practices would not have equivalent implications on the big farm and the small farm. The other area in which the narrow and broad approaches to agroecology might reach opposite conclusions is in the analysis of sustainability. A farming system that could be carried out indefinitely without harm from a purely agronomic viewpoint might nonetheless be unsustainable when analyzed in its social context.

Agroecology and Deep Ecology

Agroecology, in both the narrow and broad senses, presents itself as an objective science. Its goal is to describe the mechanisms by which the agroecosystem operates, without imposing a value judgment on the findings. But the pretense of objectivity is not valid; merely by the questions that they choose to ask, the students of an agroecosystem are smuggling values into their studies, and most often these values are utilitarian. The researchers are interested in manipulating the system to increase its utility to humans, for example by eliminating troublesome pest species, or increasing yield, or increasing efficiency.

Deep ecology is an environmental philosophy that advocates the rights and values of all species regardless of their utility to human enterprises. It promotes a program of radical restructuring and simplification of human life in deference to sustained health of all ecosystems. While it encompasses the subject matter of agroecology, it does so in a context of explicit values, without pretension to objectivity. My formal education and early employment were as a naturalist in the tropics; I feel a particular congeniality with all sorts of organisms, and I embrace the premises of deep ecology.

It is often supposed that if a person does not adhere to a religion, then they are a secular humanist. I am neither religious, nor secular, nor particularly a humanist. Humans, at their best, are a wonderful species, but this is uncommon; the majority of adults, especially in groups, are multiply flawed and often very discouraging. On the farm there are a few animals that I might rate above humans in a scale of excellence: the black-tailed hare, a superb athlete who carries with dignity his absurd ears; barn swallows, who fly so adeptly and joyfully; and the little burrowing spiders that run nimbly about among the weeds. I mention this because it bears on how I go about operating my farm, at times following a course that appears irrational and unprofitable—at odds with the common utilitarian paradigm.

Method of the Work

The method of the work is based on close observation and informal experimentation. Observation is not so simple as it might seem; humans tend to see what they expect to see rather than what is really there, and vigilance is required to avoid this trap. The first requirement for close observation is to be fully attentive; the farmer with a wire dangling from his ear, connected to an electronic device, will be a poor observer of his farm. Often some intervention is appropriate. Open your pocket knife and chip off some flakes of bark from a tree to see what's going on underneath, or slice open some developing fruits to check for color and texture and aroma, and to scout for fly larvae. I find that digging a hole in the soil with a sharp spade is always informative. You can check soil structure, soil moisture, the condition of roots, the presence of fungal mycelia, and the activities of soil organisms.

Another requirement for good observation is to be unhurried. The pace of animal life, especially of invertebrates, tends to be slow, and sometimes you have to sit and watch for a while (twenty minutes is not wasted) to see what is going on. Make rounds at midnight with a flashlight; you may find earwigs at work on your crops, a fox studying the construction of the hen house, an opossum in the fig tree, and a rat chewing on the corner of the seed cabinet. The farmer who never goes out at night misses a lot. Finally, a good education in botany, zoology, chemistry, and agronomy provides an intellectual structure into which observations can be fitted in a useful way.

Experimentation is another means to understanding. How does the watermelon crop fare when grown with one irrigation versus three irrigations? Well, try a few rows of each, and see. By dividing my planting of tetraploid anemones between two soil types, I find that the flowers grow taller in the sandy loam of the creek, but the plants in the silty clay loam of the field come into flower a good ten days earlier. Based on this experiment, I now divide my planting between the two locations in order to extend the season. This informal experimentation is a poor cousin of the formal experiments conducted at universities, which likely involve randomized block design, blinded data collection, and rigorous quantification and statistical analysis. My experiments are not randomized, the observations are not blinded, and I estimate rather than measure the quantitative variables. An account of these informal experiments would be rejected by a scholarly journal, but that does not mean that the conclusions are invalid—only that they fail to meet a particular arbitrary standard. And if the differences between two groups are so subtle that they can be inferred only from large experiments and complex analysis, then probably they're not important.

Chapter 2

The Setting

A Swainson's hawk with whom I am acquainted arrives each year in spring from his wintering ground in Argentina and takes up residence in the top of the tallest eucalyptus tree on the farm. Some years he brings a mate and they raise a pair of chicks; other years he is a bachelor.

The hawk has learned that when I am driving a tractor—plowing or mowing—small creatures are often flushed from the tractor's path and make for an easily caught lunch. And so he glides down from his high perch and settles on a fence post or a flimsy branch of a nearby tree where he can keep a close eye on my work. Whether he considers man-on-tractor to be a different species of animal than man-walking I do not know, but he is fearless of man-on-tractor. I can drive within two meters of him, seeing clearly his fierce countenance, and he does not fly away.

Imagine the final days of the hawk's journey from Argentina. Cresting the Tehachapi mountains, he finds before him the Central Valley of California, 70 miles wide and 430 miles long. The southern third of the valley, the Tulare Basin, is fed by several rivers from the Sierra to the east, but it has no alluvial exit. Whatever water enters this basin is used for irrigation; the portion not taken up by plants either enters the groundwater or evaporates. The central third of the valley, the San Joaquin Valley, is drained by the north-flowing San Joaquin River, while the northern third, the

Sacramento Valley, is drained by the south-flowing Sacramento River. The Sacramento River and the San Joaquin River join in the region known as the Delta where they turn westward, flowing into San Francisco Bay.

As the hawk flies northward from the southern end of the valley, he will be traveling along a gradient of increasing moisture. The Tulare Basin is a sparsely irrigated desert, and any unirrigated land is dusty and hostile to life. Heading north, one encounters increasing river flows and increasingly lush landscape, arriving, in

Topographical map of California, with the farm indicated by a star.

the Sacramento Valley, at formerly swampy regions now devoted to the cultivation of rice.

Some forty miles north of the Delta, the hawk recognizes landmarks of his summer residence. These would include a strip of riparian forest on either side of a small river (Putah Creek) extending eastward from the coastal mountains across the valley to the floodplain of the Sacramento River, and a row of exceptionally tall (120 feet plus) eucalyptus trees. From his cruising altitude he glides down toward the eucalyptus trees, angles his wings, and makes a neat landing on a dead branch—home for the summer.

Aerial View of the Valley

From on high, the valley is a grid of rectangular fields, with the lines of division running east/west and north/south. But here and there one sees inclusions where the property lines run at an angle other than the cardinal directions, and parcels of irregular shape may be fitted in. These are former Mexican land grants, established prior to 1850 when California became a state. The surveys of the land grants were made on the ground. The boundary might run from a notable tree to a spot of high ground, then skirt a swampy area, thence turning toward a well-defined peak in the mountains for 1,000 paces, at which point it turns NNW and runs along a ridge to a group of three oak trees, and so forth, following natural features of the land until it returned to its point of origin. In these cases, the map was a reflection of the geography of the area mapped. But after California became a state, the township and range survey, with its lines in the cardinal directions, was adopted. The township and range survey was an abstraction, created on paper in an office by someone who had never seen the territory being mapped, and had no interest in it. And when measured out on the ground, the division of land created in the survey office was often deeply illogical.

My farm was included in a Mexican land grant (Rancho Rio de los Putos), which accounts for its irregular shape (a trapezoid), and the orientation of the boundary lines, none of which is in a cardinal direction.

The Native Vegetation of the Sacramento Valley

For an ancestor of the currently resident hawk flying up the valley a few hundred years ago, not a straight line was to be seen in the valley's landscape. Although there was a sizable human population in the valley, the residents lacked tools and draft animals, nor had they the motivation to pursue farming rather than hunting and gathering. And so the appearance of the valley was not much altered from its prehistoric condition.

Excepting a few unusual features, such as the volcanic outcrop of the Sutter Buttes, the Sacramento Valley was a mosaic of oak savanna and wetlands. The oak savannas were open and parklike, and were easily traversed by a man on foot or on horseback. Widely spaced valley oaks (*Quercus lobata*), many of them wider than tall and of prodigious size, were scattered over a prairie of perennial grasses. There was no understory of shrubs except along the edges of the rivers. The wetlands comprised seasonally flooded marshes of which the dominant plant species was the tule (*Scirpus acutus*).

Remnants of both oak savannas and tule marshes are still to be found. Within a mile of my farm is a slough where tules still grow abundantly, and in the other direction, also within a mile, is a regenerating oak savanna in what had been a well-run kiwifruit vineyard. But the kiwifruit market failed and the vineyard was abandoned. Almost immediately native oaks began to appear in the vineyard, the acorns planted by scrub jays and squirrels. A decade after the vineyard was abandoned, the trellising—stout poles of pressure-treated lumber and miles of high-strength trellising wire—was pulled out and hauled away. Without irrigation, the kiwi vines died out, and the Eurasian weeds that formed the understory of the oaks were increasingly replaced by perennial grasses. The plot is well on its way to becoming an oak savanna of the sort it was originally.

Although the land on my farm was first cleared 150 years ago and has been farmed continuously since then, the resilience of the oak savanna is still everywhere evident. Valley oak seedlings sprout on any unplowed ground, especially in the orchards, and I have to keep after them with a sharp spade or they will overtake the cultivated trees. Around the house I have allowed a dozen or more volunteer oaks to persist, and in a few decades they have become substantial trees, sixty feet tall and with the diameter of the main trunk reaching two feet. Perennial grasses have not been as aggressive, but I suspect that if the farm were abandoned, they would eventually return.

The Size of My Farm

The area of my farm is 21.33 acres (8.65 hectares). The measure of an acre arose in England a few hundred years ago, and it was based on the average size of a parcel of land that an ox could plow in one day. This was usually configured as a length of one furlong (660 feet—the length of furrow that an ox could plow before it had to be rested) by one-tenth of a furlong (66 feet). The reason for the long skinny plot is that it is difficult to turn an ox and plow around, and so it makes sense to spend more time plowing and less time turning. The length of 66 feet, one-tenth of a furlong, is called a "chain." A mile is 80 chains, and an acre is 10 square chains. A quarter of a chain, 16.5 feet, is a "rod." This sounds archaic, but when I was a child it was still common to hear older people describe a distance in rods, and my arithmetic lessons in grade school included calculations of distance in chains, rods, and furlongs. Many rural road easements are one chain (66 feet), and field fencing is still sold in rolls of either 165 feet (10 rods) or 330 feet (5 chains).

The extent to which an acre is a useful measure depends on the nature of your enterprise. In Burma and Thailand a more common measure is the *rai* (0.4 acres), which is approximately the area of rice land that a water buffalo can plow in a day. In China land is traditionally measured by a unit of *mu* (one-sixth of an acre), and in Japan, where much of agriculture is based on intensive hand labor, measures of land area include the *tsubo* (36 square feet) and the *tan* (0.25 acres). For much of what I do, an acre is too big of a measure to be useful to me, and instead I measure land area by beds. For me, a bed is four feet by two hundred feet. And so as I make my plans I think in terms of two beds of lettuce, and twelve beds of watermelons and twenty-six beds of sunflowers planted at one-week intervals starting in February. In my orchards I find it more useful to count trees (60 apricot trees, 1,700 olive trees) rather than to calculate area.

At the other extreme, land in the Corn Belt of the United States is often measured in quarter sections (160 acres) or sections (640 acres). This reflects the scale of industrialized grain farming, as well as linguistic convenience: it's easier to say "two quarters" (three syllables) than "three hundred and twenty acres" (eight syllables).

The average size of a farm in the US in the 2014 agricultural census was 234 acres of cropland. The median size was only 45 acres. My farm is well into the bottom quintile of farm size, which classifies it as a small farm. One hundred fifty years ago it would have been called a "market garden," and in other countries today it would be called a "smallholding"; by local custom it is called a "small farm." More pertinent than the number of acres is the potential gross income and whether the farm qualifies as an economic unit that can sustain a family. In terms of gross income, my farm is above the 80th percentile, and clearly is a valid economic unit despite its small size.

Topography

The Central Valley of California was at one time an inland sea, and like most formerly submarine lands, it is very flat. In my district, the average slope of the land is ten inches in a thousand feet (0.08%), sloping to the south-southeast. This is only just enough slope to get water to flow down an irrigation furrow. Atypically, my farm has some topography to it because of the creek. During low water seasons, the level of the creek is about 30 feet below the level of the land; from the creek the land slopes upward, forming several benches, one of which, an acre in size, I farm. At the edge of the creek channel, there is a long berm a few rods in width that is elevated about 30 inches above the surrounding country. Most of my buildings (studio, house, guest house, workshop, barn, greenhouse) are lined up on this elongated "hill." People from mountainous country would find it ludicrous to call a rise of 30 inches a hill, but in flat country it is significant. In historic floods, this was high ground that stayed dry.

Layout of the Farm

The farm may be divided into four roughly parallel zones. About five acres along the creek is covered with native riparian forest—willows, cottonwoods, oaks, walnuts, and box-elders, with an understory of wild roses, blackberries, native grapes, and formerly, poison oak (my neighbor and I eradicated poison oak on our pieces). The native forest is nearly free of invasive species (arundo, tamarisk, eucalyptus) that degrade other riparian lands in the valley, and because the channel of the creek is fairly wide (600 feet) at this location, the Army Corps of Engineers has refrained from bulldozing it into a ditch.

Coming up from the creek channel to the higher ground, one first encounters the row of buildings, as well as a line of eucalyptus trees planted in the 1880s. These trees are huge, dangerous, and senescent, and I would expect that not much will be left of them twenty years from now. There are also two rows of citrus trees planted on the higher ground with the idea that, during a freeze, cold air would drain away from them.

Continuing south from the line of buildings is a region of open ground planted to annual crops or herbaceous perennial crops. This is divided into seven fields, ranging in size from 0.5 to 1.5 acres, the fields delineated by hedgerows. In addition, a third of an acre is planted to honeylocust trees providing shade for some shade-loving crops (gerbera, as well as nursery stock).

The remaining area, which comprises the majority of the farm, is planted to orchards: apricots, figs, olives, persimmons, and citrus, with olives taking up about 85 percent of the orchard ground.

The surrounding lands are planted as follows: to the east, walnuts; to the south, walnuts; to the west, walnuts; and to the north, across the creek, tomatoes rotated with wheat. The trend of the last two decades in this general region has been that land formerly planted to field crops and row crops is being steadily converted to orchards (walnuts, almonds, pistachios, and plums), which are more dependable and more profitable.

Failure of Alley Cropping

The current layout of my farm is not my original vision. When I started out I proceeded with the idea of parallel rows of trees, the rows fifty or sixty feet apart, with herbaceous annual or perennial crops in the intervening alleys. The trees, being widely spaced between rows, would be able to send roots without competition into a large volume of soil from which they could obtain water during the dry season. And the herbaceous crops, being shallow-rooted, would not interfere with the extensive root systems of the trees. In addition, the trees would serve as windbreaks.

As it turned out, I had difficulty establishing the herbaceous crops in the alleys because of predation by birds (sparrows, quail) and rodents. Many small birds and rodents will not venture into a large open area that lacks cover to which they might quickly flee if a predator appears, and so they confine their foraging to areas close to trees. Ground squirrels, a troublesome pest, prefer to build their burrows at the base of a tree. And so the alley-cropping design provided perfect habitat for these predators of my crops. As I came to understand this, I eventually had to admit failure. I pulled out most of the tree rows to reconfigure the farm so that larger open areas were separated from the tree crops. The trees I was growing (bay, incense cedar, *Prunus*) were mostly bushy and dense close to the ground; had I chosen more open trees with a clear bole I might have had a better outcome.

There is one remnant of the alley-cropping concept still in use. Where young olive trees are planted (often only knee-high at planting time), I grow watermelons in the alleys. The watermelon vine and foliage are bitter and are immune to predation, and so the alley situation works well. As the olive trees get up to size, the alleys become too shady for melons.

Spacing of Trees

The trend of orchard horticulture over the last eighty years has been toward smaller trees planted closer together. This is usually achieved by a combination of dwarfing rootstocks and heavy pruning. For example, Roeding (1919) recommends a spacing of 40 feet by 40 feet for olive trees (27 trees per acre); now many olives are planted at a spacing of 5 feet by 10 feet (870 trees per acre). The advantages of higher-density planting are more rapid maturation of the newly planted crop, easier harvest, and less disruption if a tree should die. From the nurseryman's point of view, high-density orchards allow him to sell a lot more trees, and so he also promotes this approach more enthusiastically than it might deserve.

My apricots are planted at a spacing of 20 by 20 feet, farther apart than is now common. This spacing provides each tree with enough soil volume for its roots so that this, augmented by drip irrigation (which is limited to the tree row), provides sufficient moisture for the trees to survive the dry season and bear an average crop. If I were replanting apricots I would stick with that spacing, which is appropriate to my conditions.

The first block of olives I planted were spaced at 11 by 18 feet, which was a commonly used spacing at the time. But I came to feel that the trees were too crowded, and in the second block I increased spacing to 12 by 22 feet. A subsequent block is planted at 12 by 25 feet, a spacing that makes for happier trees. For the closer spacing, keeping a balance between the volume of soil available to the root system and the size of the aboveground portion of the tree requires severe pruning every year. The closer spacing also demands a more scrupulous attention to irrigation. In addition to allowing the trees a greater volume of soil, the wider spacing provides adequate room for the operation of the mechanical harvesting machines, should I choose to use them, which might not be able to maneuver in the tighter rows.

I have considered removing every second row of the closely spaced olive trees, leading to a spacing of 11 by 36 feet, or removing every second tree in the rows, resulting in a spacing of 22 by 18 feet. With less pruning required and a greater soil volume available to the trees, I believe that the yield would not be greatly reduced. Another possibility is to cut off every second tree at ground level and allow it to sprout from the stump (alternate tree coppicing). The trees are on their own roots and not grafted, so there is no danger of winding up with an undesired rootstock. After a period of years, as the coppiced trees grow to full size, the alternate trees could then be coppiced. These are possibilities that I am thinking about, but to which I have not yet committed. I have tried cutting off a few olive trees at ground level to measure the labor required and the speed with which the tree returns to bearing size.

If I had been thinking clearly about creating a farming system similar to the native vegetation—a savanna with tree canopy covering about 50 percent of the ground and the roots of the trees spreading through 100 percent of the ground—I would have adopted a wider spacing of trees right from the start. I suspect that the native savanna, with trees having root systems far larger than their canopies, constitutes an optimization of growth in a water-limited system.

Chapter 3

Crops

The region of my farm is especially favored for agriculture. The soils are fertile, stone-free, and deep. Rainfall in the watershed is sufficient that irrigated farming can be carried out without depleting geologic groundwater. This is because about three-quarters of the watershed is mountainous and is not farmed, and runoff from that area serves to irrigate the remaining arable quarter. The climate is cold enough to provide adequate chilling for stone fruits, yet warm enough that citrus can be profitably grown. Dryness of the air in the summer season makes the region inhospitable to plant diseases, which are much less prevalent than in more humid zones. Absence of clouds throughout much of the year guarantees high light intensity that supports rapid plant growth. The land is so flat that furrow irrigation is feasible. Rainfall is usually gentle, which together with the flatness of the land means that the risk of soil erosion is essentially zero; indeed, historically this is a zone of soil deposition. Farming practices that would be ecological felonies in other regions, such as leaving the earth plowed bare through the rainy season, may here be practiced seemingly without penalty. More than a century ago, Professor Hilgard remarked, "Any damn fool can be a good farmer on that land."

The social and institutional side of farming is also favorable. The region has a fairly stable and orderly society, and the rule of law predominates most of the time. Infrastructure of roads and irrigation systems are well designed and well maintained. Dependable markets operate for both purchase of equipment and supplies and sale of crops and livestock. The local populace of customers is educated, enthusiastic about good food, and willing to pay a premium for exceptional produce. Banking and finance are attuned to the needs of farmers. And the presence of a major agricultural university (UC Davis) with a knowledgeable and accessible staff spreads a benign influence throughout the district.

On the negative side, retail prices of produce are much lower than they are in other parts of the country. A premium bouquet of flowers that I sell for $7 would fetch $30 in the northeastern states. Top-grade organic apricots that I sell locally at $2 per pound retail would sell at $4 or $5 per pound in San Francisco, and considerably more in distant markets. This is simply a classic example of the actions of supply and demand on price.

Choice of Crops

In choosing which crops to grow, there is a compelling logic to growing crops that are already widely grown in the region. These have proven themselves adapted to the soils, climate, and markets, and supporting technology and information are readily available. My district has a diversified agriculture, in which the principal crops are the following: wheat, safflower, sunflowers, tomatoes, peppers, cucumbers, melons, squash, beans, onions, garlic, corn, hay, wine grapes, table grapes, walnuts, almonds, pistachios, chestnuts, olives, plums, apricots, peaches, citrus fruits, berries, and figs. Since it was first farmed in the 1860s, my land has been planted to hay, pasture, sunflowers, wheat, grapes, oranges (killed in the great freeze of December 1932), safflower, barley, nursery stock, and melons. Evaluating several of these illustrates some of the constraints of a small farm.

I could easily till a field and plant wheat in the fall, which would grow over the winter with rainfall as the moisture source. I would hire a custom harvest crew to harvest it, and sell it to a commodity broker. This is an easy and trouble-free crop. The problem with it is that it has very low value per acre, and the income would hardly pay my property tax, let alone give me a living. So for a farm as small as this, I need a crop with a high value per acre. The highest values per acre are to be found with crops that are labor-intensive: cut flowers, berries, stone fruits, and fresh market vegetables. Any crop, such as wheat, whose production is fully mechanized will be profitable only on a large scale.

Apricots are a high-value, labor-intensive crop, and grow well here. The problem with apricots is that even spreading out the season with several varieties, the harvest is fast and intense. If I planted all the land to apricots, I would have to hire a crew of 40 or 50 workers to get the fruit harvested—three weeks of pandemonium, followed by 48 weeks of nothing much to do beyond some irrigation and pruning. This is a second issue that has to be resolved—how to spread my labor out through the year so that I always have work to do. This is best achieved by growing a diversity of crops, rather than just one crop.

A third constraint derives from finding a reliable market. For example, there are a number of farms in the area with intensive production of heirloom tomatoes. This market is already saturated at the farmers markets, fruit stands, and independent grocers; it would be difficult to compete in a market like this that lacks marginal demand.

Another constraint is self-imposed: I don't want to grow crops (e.g., almonds) that are highly dependent on honey bees for pollination. Bee culture in California has had many problems during the last few decades, and rental of hives for pollination is expensive and sometimes a sufficient number of hives cannot be found. I prefer not to have that worry. I have also limited myself to drought-tolerant crops, recognizing that abundant water may not always be available.

Finally, I believe it's important to grow crops that you like. There are a number of things that I could grow—rutabagas, turnips, kale, Christmas trees, cabbage—that I'm not interested in, and because of that I would do a less good job than I would for crops about which I am enthusiastic.

Crops Grown on the Farm

Table 3.1 lists the crops grown on the farm, with the number of cultivars for each species, and the approximate number of plants. In addition to the crops on the list, we keep a garden for our own use, the produce of which is not for sale. The garden includes onions, leeks, potatoes, sweet potatoes, beets, chard, broccoli, peas, beans, tomatoes, peppers, aubergines, basil, and squash. I also grow green manure crops, planted each autumn and sustained by winter rains; they are plowed down in the spring as conditions become drier. These include clovers, sub-clovers, medics, sweet clover, fava beans, peas, oats, wheat, rye, radishes, turnips, and mustard. And finally, I grow a great many sorts of plants of no known utility simply out of interest in them.

Table 3.1 includes 204 cultivars represented by more than 50,000 plants. This is not just a randomly selected jumble of crops, however. It is the product of three decades of evolution, during which time I have figured out exactly what to grow, when to plant it, and how much to plant, in order to have a steady production that will exactly meet the demands of the market. For example, I make 26 separate plantings of sunflowers for bunching at approximately one-week intervals, starting in February, with the cultivars changing throughout the season to accommodate different adaptations to day length. When I shift from 'Pro-cut' to 'Superior Sunset' in the first week of July, I make a planting of each on the same day; 'Superior Sunset' is a 58-day crop as opposed to 50 days for 'Pro-cut,' and so the harvests will be a week apart despite same-day planting. Other sunflower cultivars, which produce smaller sunflowers on a branching plant that are useful for bouquets, are sown on a different schedule. This is not

Table 3.1. Crops Grown on the Farm

FRUIT & VEGETABLE CROPS

Type	# of Cultivars	# of Plants
Olives	21	1,700
Apricots	2	60
Figs	13	48
Persimmons	2	17
Oranges	4	14
Lemons	2	9
Bergamot	1	18
Yuzu	1	4
Kumquats	2	16
Mandarinquats	1	10
Other citrus	6	9
Damsons	1	5
Nectarines	3	4
Plums	3	3
Quince	2	16
Apples	2	2
Pomegranates	2	7
Blackberries	4	390
Grapes	5	16
Watermelons	2	900
Cantaloupes	2	860
Cucumbers	1	700
Lettuce	1	400

FLOWER & FOLIAGE CROPS

Type	# of Cultivars	# of Plants
Tulips	16	18,000
Anemones	1	1,000
Ranunculus	5	5,000
Dutch iris	1	2,000
Peonies	12	1,455
Tuberose	1	1,980
Alstroemeria	1	400
Crinadonna	1	360
Safflower	1	110
Leucojum	1	180
Lycoris	1	60
Rosemary	2	70
Lavender	2	145
Statice	4	80
Sunflowers	5	11,400
Celosia	3	340
Scabiosa	1	400
Echinops	1	180
Gerbera	32	2,160
Gomphrena	3	155
Marigolds	2	200
Molucella	1	130
Sweet peas	8	330
Lilac	2	120
Prunus mume	2	42
Godetia	2	400
Incense cedar	1	52
Myrtus	1	6
Bay	1	66

a scheme I figured out in just one year; it was arrived at following years of observation.

Even longer than the list of cultivars in table 3.1 is the list of plants that I have grown as a trial and rejected. This includes some major plantings (800 vines of table grapes; 180 Clementine mandarins; 98 fig trees that were killed by gophers) that simply didn't work out, as well as numerous smaller trials of herbaceous plants, including hundreds of flower crops. If it's listed in a seed or bulb catalogue somewhere, I've grown it.

The only vegetable that I grow for sale is lettuce; my other crops are fruits or flowers. Vegetable production requires the least amount of capital and has the fastest returns—within 30 days you can have a product to sell. And so vegetable growing attracts the most severely undercapitalized farmers, who also are often the most desperate, with the result that this is a very price-competitive market. We stay away from it.

The number of plants—about 50,000—becomes meaningful when you consider that all of my crops are hand-harvested. A wheat field has far more than 50,000 plants per acre; that acre is harvested by a diesel-powered combine in a matter of a few minutes, and the individuality of the plants is hardly noticed. But my operation requires recognizing each plant as an individual, whether it is an olive tree or a watermelon vine or a tulip. When farming on a small scale, it is the labor-intensive aspect of the process that gives the product its value, and because my competitors in the marketplace are also constrained to hand-harvest, we are on an even basis. That is, I am not competing with a machine—a contest that I would surely lose—but only with others whose circumstances are similar to mine.

In addition to allowing me to spread my labor throughout the year, the diversity of crops also mitigates the risk of crop failure. In any given year, somewhere between 10 and 40 percent (occasionally even 60 percent) of my plantings fail; to look at it positively, I always have between 60 and 90 percent success. If I grew only one crop and it failed, the consequences would be devastating.

Another advantage of a diversity of crops is that my labor is highly varied. Olive harvest and olive tree pruning are the only work I do that runs for 10 hours straight each day. The rest of the year, my day is broken up into a variety of different chores, which is physically and mentally more enjoyable than doing only one thing.

The Annual Cycle

In most temperate regions there is an annual cycle to growing crops: planting in the spring, cultivating through the summer, harvest concentrated in a few months of late summer and fall, and rest in the winter (with any luck in a subtropical place). That cycle doesn't apply in my district. With my array of crops I am harvesting fruit twelve months of the year and flowers eleven months (see "Monthly Harvest of Fruit" sidebar). I am also planting every month of the year.

This continuous production is integral to our marketing. Our local farmers market, which is where we sell nearly all that we produce, is held twice a week, all year. And so we are able to keep our market income flowing throughout the year, and are spared the discomfort of husbanding a steadily shrinking bank account over a long period with no income.

Another aspect of maintaining a steady income is converting a perishable product, like fresh fruit, into a non-perishable product, like jam. For example, apricot harvest is concentrated in a hectic period of twenty days. We sell fresh apricots at the market, but so do a dozen other farmers at precisely the same time, so the supply far exceeds the demand. But we also make the majority of the harvest into apricot jam, which we can then sell throughout the eleven months until the next crop comes in. Not only does this prolong the opportunity for sales, but it also significantly increases the value of the crop: two dollars worth of apricots becomes seven dollars worth of jam.

The majority of fresh fruit on the farm is made into jam (plum, damson, fig, apricot, blackberry, orange, yuzu, bergamot, lime, and lemon). Ninety-nine percent of the olives are processed for olive

Monthly Harvest of Fruit

JANUARY

Oranges
Lemons
Limes
Bergamot
Kumquats
Yuzu
Olives

FEBRUARY

Oranges
Lemons
Kumquats
Bergamot
Olives

MARCH

Oranges
Lemons
Mandarinquats
Olives

APRIL

Oranges
Lemons
Mandarinquats

MAY

Grapefruit
Lemons
Blackberries

JUNE

Apricots
Plums
Blackberries
Grapefruit
Figs
Cucumbers

JULY

Grapefruit
Lemons
Plums
Nectarines
Apples
Watermelons
Cucumbers
Cantaloupes

AUGUST

Grapefruit
Lemons
Plums
Apples
Figs
Grapes
Watermelons
Cantaloupes
Cucumbers

SEPTEMBER

Figs
Lemons
Damsons
Apples
Watermelons
Cantaloupes
Cucumbers

OCTOBER

Persimmons
Quince
Olives
Pomegranates
Apples

NOVEMBER

Persimmons
Pomegranates
Quince
Olives
Mandarins
Lemons
Yuzu
Apples

DECEMBER

Oranges
Kumquats
Mandarins
Persimmons
Olives
Apples

oil. Excess olive oil is made into soap, or mixed with beeswax from the farm to make hand salve and lip balm. Our hot, dry climate is also conducive to preserving fruit by drying. We dry figs, apricots, nectarines, plums, tomatoes, and olives (packed in salt); these are primarily for our own use rather than for sale. To dry them, we simply split open the fruits and lay them on racks in the sun.

Sources of Plant Materials

We obtain plant materials wherever we can. Many of the trees and other perennial plants have come from specialist nurseries, or from the California Department of Forestry. Olive trees, which are not grafted but grown on their own roots, we propagate from cuttings. Many starts have come from friends. Many of the fig cultivars came from germplasm collections of the University of California, Davis and Riverside campuses. As with olives, we propagate figs, pomegranates, grapes, and plums on the farm from cuttings. Some crops (peonies, blackberries) we increase by divisions. Our citrus trees are grafted on a dwarfing rootstock by a local nursery, except for yuzu, which I grew from seed.

The bulbs that we plant originate for the most part in Holland or Israel. Our climate is not really appropriate for tulips, and so we purchase bulbs from a broker who stores them for us at 5 degrees Celsius (41°F) to satisfy their chilling requirements before we plant them. The tulips are treated as annuals, and are plowed down after harvest.

Seeds are purchased from a variety of sources, but I also save seeds whenever it's feasible. The sunflowers that I grow are sterile varieties, and so I cannot save seed, and must buy it each year. But for most other annual crops, including crops that are listed as F1 hybrids, I let some plants go to maturity and collect seed from them. Some seeds I obtain from my local supermarket where they are being sold as food; these include some of the grains, as well as peppers, tomatoes, and melons from which I've saved seeds.

If you look through wholesale seed catalogs, you will find relatively few open-pollinated cultivars. Almost all listed cultivars are either F1 hybrids or are vegetative propagules from cuttings or tissue culture. Two advantages are claimed for F1 hybrids: increased vigor, which is dramatic in a few cases but mostly not a factor; and increased uniformity of size, appearance, and timing of maturation. Increased uniformity is critical for machine-harvested crops and for crops sold and processed through the industrial food system. But on a farm such as mine, uniformity is not particularly advantageous. I prefer variation in maturity, so that I have a continuous harvest rather than a boom-and-bust harvest cycle. And if, for example, there is variation in the size of watermelons I grow, that suits my market, as some customers prefer a smaller melon and some prefer a bigger one.

From the breeder's perspective, there is another overwhelming advantage to F1 hybrids: the farmer can't save seed to replant the following year, because the crop won't come true. Moreover, the seed company enjoys a monopoly because competitor seed companies lack access to the parent lines of the hybrids. This forces the grower to buy seed every year from the breeder, creating a situation that is reflected in very high prices for hybrid seed compared to open-pollinated seed. For example, a standard, open-pollinated cucumber, 'Marketmore 76', sells for less than 2 cents per seed, while an F1 hybrid cucumber, 'Tasty Jade', sells for about 45 cents per seed, and some of the varieties bred for glasshouse culture cost about 80 cents per seed.

In school we were heavily indoctrinated with the idea that F1 hybrids won't come true from seed, but I regularly save seeds from hybrid crops and find that the F2 and subsequent generations do not show the chaotic variability that we were told to expect, but rather, are nearly identical to the F1's. Perhaps the hybrids were derived from two inbred lines that differed by only a single gene, or maybe a gene whose expression is invisible, such as a trait for disease resistance. But it is also possible that these are simply inbred lines that are dishonestly described as F1 hybrids in order to raise prices and boost seed sales.

Weeds

Weeds are an unwelcome but inevitable feature of agriculture. Many weeds have coevolved with the crops that they infest, and have spread with them throughout the world. I divide the weeds on my farm into three groups. Winter weeds germinate with the autumn rains and flourish through the winter and spring. The most widespread include foxtails (*Hordeum leporinum*), wild oats (*Avena fatua*), ripgut (*Bromus rigidus*), wild radish (*Raphanus sativus*), and mallow (*Malva parviflora*). The summer weeds germinate late in spring and flourish through the summer months. Most troublesome of these are pigweed (*Amaranthus retroflexus*), wild lettuce (*Lactuca serriola*), and the most troublesome weed of all, field bindweed (*Convolvulus arvensis*). In addition to the weeds mentioned, there are dozens of other weed species that either are not widespread or show up only every few years.

The third group of weeds are what I call "zero-tolerance weeds," which is to say that these are weeds that I do not permit on the farm. If I see one, I stop whatever I am doing, including driving a tractor, dig up the weed, put it in a bag, and get it off the farm. There are four weeds in this group: Bermuda grass (*Cynodon dactylon*), nutsedge (*Cyperus esculentus*), yellow star thistle (*Centaurea solistialis*), and puncture vine (*Tribulus terrestris*). What puts a weed into the zero-tolerance group is that it is a weed which, once established, would be very damaging to the farm operation and difficult to control. Finally I should mention several tree species (valley oak, coast live oak, olive) that germinate frequently and abundantly throughout the farm, and could be classified as weeds.

The act of cultivating the soil and the act of planting weeds are the same act. Cultivation may scarify hard seed coats and expose the seeds to light, both of which stimulate germination.

For perennial weeds like field bindweed or Bermuda grass, cultivation cuts up rhizomes and roots into small pieces, each of which is capable of growing and forming a new plant. This concept is borne out by looking at the weed flora of the orchards. In the apricots, where the orchard floor is not cultivated but only mowed, or some years not even mowed, the understory forms a stand of wild oats with some sub-clovers and medics. If I mow in the early spring, instead of wild oats I will get foxtails. Bindweed, mallow, and other weeds that were present before the apricots were planted, have disappeared in the absence of cultivation.

The older portions of the olive grove have essentially no weeds. Some medics (especially *Medicago polymorpha*) and sub-clovers that I planted years ago germinate with fall rains and go to seed by early spring. There is no field bindweed, pigweed, or mallow, and only a few grasses. There are several reasons for this. The olives are evergreen trees that cast a dense shade. The branches pruned from the olive trees are chipped on site and form a woody mulch. The olive pomace that is the waste from olive oil extraction is spread back into the grove as fertilizer; it is mildly phytotoxic, and prevents growth of weeds. And finally. because olive harvest occurs during the rainy season, tractors are driven on damp ground and the soil surface has become compacted in a way that is inhospitable to germination of weed seeds.

The upshot is that for the established tree crops, weeds are not an issue. It is in the annual herbaceous crops grown on open ground that weed management is a problem. Since I do not use chemical herbicides, I must use a suite of complementary cultural practices to keep weeds under control. A heavy planting of winter green manure crops (oats, vetch, bell beans, peas, and mustard) can crowd out winter weeds. However, many of my annual crops are grown over the winter, and the green manure crop cannot be grown among them. The best result is to plant winter crops as a rotation on ground that had a winter cover crop the previous year, or, better yet, previous several years. A related practice is to allow an area of land to be fallowed in grasses (this usually ends up being foxtails) for four or five years or more. This builds up a deep sod which, when returned to cultivation, has many fewer weeds (except, of course, for foxtails). I am limited in my ability to use long cycles of grass fallow or green manure crops because I don't have enough land. Ideally, I would keep most of my land for annual crops in a grass-and-clover fallow, the portions of which used for crops would be disturbed only every fifth or sixth year. This was essentially the medieval agriculture of Europe; it is a very effective, low-input approach to soil management.

For fall planting before rains have started, and for summer planting, I find it advantageous to work up the beds and irrigate them, allowing the weeds to germinate, which they usually do rapidly. The weed seedlings are then burned off with a propane torch before the transplants or seeds are set out. Bulbs (tulips, ranunculus, anemones), which will take a few weeks after planting to start growing, can be planted, the beds irrigated, and the weeds burned off before the crop emerges.

Herbaceous perennials, such as peonies and blackberries, are supplied with a mulch of wood chips to mechanically suppress weeds (although bindweed will come up through mulch even if it is one meter thick!). I can get baled rice straw locally and spread it as a dense mulch for about the same cost as biodegradable paper mulch. This has the advantage of building soil organic matter when it is plowed in at the end of the season.

Crops that are not mulched can be cultivated by a tractor pulling a cultivating sled fitted with sweeps and knives or by a dedicated cultivating tractor. It is difficult to get close to the crop this way, but weeds in the furrows can be cleaned out. Ultimately, I depend a great deal on getting out into the field with a hoe and chopping out the weeds one by one. Manuals of farming warn against hand-weeding; the labor cost is so high that it quickly consumes any profit that might be had from the crop. But since I have no employees and am doing the weeding myself, the cost of labor is not an issue, and I enjoy weeding with a hoe. In summer I usually spend the few hours after the evening meal and before darkness weeding—a relaxing end to the day.

In the 18th century, Gilbert White (*The Natural History of Selbourne*, 1789) managed weeds on his seven-acre English farm by employing a full-time weeder woman. Peter Henderson, a New Jersey market grower for the New York market, dismissed the subject of weeds in a single paragraph in his book of 1867, *Gardening for Profit*. Henderson advised hiring an Irishman for a dollar a day—one Irishman per acre is about the right number—and giving him the task of ensuring that not one weed appeared on his acre.

One additional technique of weed management should be mentioned, and that is crop abandonment. Almost every year I have crops that I didn't get to in time and that have become infested with larger weeds, and I have come to accept the idea that in this case the best thing to do is to mow the crop, disk up the ground, and just write it off as a loss.

Chapter 4

Animals on the Farm

A census of farms in my district from 130 years ago lists, for each farm, the crops grown and the livestock raised. Nearly all of the farms had horses for transportation, mules for plowing, a few milk cows, hogs, geese, ducks, and chickens. About half of the farms kept some sheep, and a few had steers. The milk cows, swine, and poultry were for the use of the farm family, with the crops sold for cash. This represents something of a midpoint in the evolution of American farms from the 18th-century situation—in which farms were subsistence enterprises with a diverse production consumed on the farm, and with cash crops for sale comprising only a small part of the farm activity—to the 21st-century situation, in which 100 percent of the products of the farm are sold in the market economy.

Swine culture has almost entirely disappeared from the region of my farm, although occasionally a 4-H student will raise a pig. A few farmers keep some poultry, and some have a kitchen garden, but the norm now is that the farm is entirely a market enterprise, and all of the farmer's food is purchased. Along with this shift to a market economy there has been an increase in specialization, so that farms raising crops do not raise livestock, and livestock ranching is a specialty in which crop production does not play a part. This is true across the country. Even for those farms that grow crops, the trend is toward increasing specialization, so that a local farm of a thousand acres is unlikely to produce more than two or three different crops, and frequently just one. The child's storybook concept of a farm, with its diverse crops integrated with a variety of animals, is no longer a reality in this region.

The Ecological Side of Animals on the Farm

Historically, sheep and goats coexisted with the horticultural activities of farming, but were not fully integrated. Dairy cows, if suitably integrated with cropping, can be a critical part of an ecologically sound farm, as was formerly the case in much of northern Europe and eastern North America before animal culture was widely divorced from crop production. Swine and poultry show the greatest ecological integration into a farm. They can feed on weeds, crop refuse, and kitchen scraps, and forage for insects and seeds. A woodlot is ideal for ranging swine and poultry much of the year, or they can be turned into fields following harvest to scavenge spilled grain and crop residues. A Hmong friend of mine who operates a market vegetable farm keeps pigs and ducks, which he feeds on vegetable trimmings and unsalable produce.

It is often stated that manure from animals on the farm can increase soil fertility, and this is given as a compelling reason for keeping livestock. While this can be true, usually it is misunderstood. Suppose you put sheep into a fenced orchard where they will feed on the understory weeds and deposit their manure. This actually decreases the fertility of the orchard. For example, if there are 100 units of nitrogen in the weeds eaten by the sheep, then some of that nitrogen will be converted to muscle and blood and wool, and some will be converted to ammonia that is lost to the atmosphere. Perhaps only 80 units of the nitrogen wind up in the manure. So there is a net loss of nitrogen compared to simply plowing down the weeds.

Consider a slightly different situation in which the sheep are pastured by day in some remote grasslands, and are penned at night in the orchard, where they defecate. In this case, fertility of the orchard is increased since nutrients from the pasture are transported in the sheep's gut to the orchard, where they are deposited. That is, livestock can transport nutrients from one place to another, and can make them more rapidly available, but they do not create nutrients. Sometimes this is referred to as an outfield/infield system; the livestock are pastured by day in the outfield, where they graze, and at night are penned in the infield where their manure increases soil fertility at the expense of the outfield.

There is another sense in which livestock improve the fertility of the farm. If the farmer needs to sell $1,000 worth of product in order to raise cash, selling animals or animal products removes fewer nutrients from the farm than the equivalent cash value of crops. For example, 300 dozen eggs contain about 9 pounds of nitrogen; the equivalent cash value of wheat (5 metric tons) contains about 200 pounds of nitrogen. So selling the animal product (eggs) conserves farm nitrogen much more than selling a crop (wheat).

Domesticated Animals on the Farm

We keep half a dozen hens and a variable number of cats. Although the hens get some purchased feed, much of their diet is kitchen scraps and products from the farm. In the summer months their feed consists mostly of watermelons and cantaloupes, which have a good balance of high-protein seeds and high-carbohydrate flesh. The hens are kept in a large pen. We are unable to let them range freely because of the high density of predators on the farm. The hens provide far more eggs than we use, and the surplus is bartered or given away.

The cats are fed commercial cat food, but primarily they live on rodents. We have one house cat who can come and go freely through a small door, and she has a habit of bringing rodents she has killed into the house to show to us before she eats them. She averages about 400 rodents per year—mostly voles and gophers, but occasionally a rat or a squirrel or even a juvenile hare. We have had this cat for 15 years; that computes to more than 5,000 rodents that she has killed, nearly all within 100 yards of the house. This gives an idea of the density of rodent populations on the farm.

Two beekeepers have hives of bees on the farm. The diversity of flower crops provides a year-round source of nectar and pollen. We are glad to have the bees for pollinating watermelons and cucumbers, and a block of beeswax and occasional jar of honey are welcome gifts. The relationship is mutually beneficial, and no rents are exchanged.

Wild Animals on the Farm

The forest along the creek that forms the north boundary of the farm is one segment of a long, very narrow strip of riparian forest that extends from the coastal mountains to the delta of the Sacramento River, and it provides a corridor along which many animals move. On my farm, the creek itself hosts beavers, river otters, and mink, as well as an abundance of fish and turtles. In late autumn Chinook salmon swim up the creek to spawn. The riparian woods are home to raccoons, opossums, skunks, grey fox, red fox, coyotes, and bobcats. On the cultivated ground there is a scattering of lizards, frogs, and snakes; gopher snakes and green racers are most common, but sometimes I get the rare treat of seeing a handsome California king snake sliding through the grass. The chief wild inhabitants of the cultivated ground are rodents. The rodent populations are dense, and include rats, mice, voles, gophers, tree squirrels, ground squirrels, rabbits, and hares.

The bird fauna is diverse and abundant, ranging from the tiniest hummingbirds to the large wild turkeys that wander the groves eating fallen olives. Much of my farm work is carried out to the accompaniment of sardonic commentary from crows. The great number of raptors (see "Birds of Prey on the Farm" sidebar) reflects the abundance of rodents on which the raptors feed.

It is important to note also the animals that are not present on the farm. We have no poisonous snakes, although rattlesnakes are to be found as close as eight miles away. We also do not have deer or wild pigs, extremely destructive pests of crops that bedevil farms just a few miles distant but that have not ventured this far into the valley.

I have attempted a rough estimate of the biomass of larger herbivores (rodents plus turkeys) on the cultivated portion of the farm: 70 squirrels (205 lbs), 800 pocket gophers (400 lbs); 1,000 voles (200 lbs); 20 hares (100 lbs); 20 rabbits (40 lbs); 50 rats (50 lbs); 200 mice (20 lbs), and 20 turkeys (140 lbs). The total, 1,165 pounds, is roughly similar to the biomass of sheep that would be stocked on this ground if it were half irrigated pasture and half dryland pasture, which is about the ratio of irrigated to unirrigated land for my farm.

In folk literature of centuries past, animals formed a vital part of the human social community. I find that to be the case for me as well. Much as I appreciate the company of animals (including insects and spiders), I have no delusions that this is reciprocal. Most of the creatures regard me with indifference or alarm. But some years ago there was an old coyote, his whiskers white (like

Birds of Prey on the Farm

Golden eagle
Red-tailed hawk
Red-shouldered hawk
Swainson's hawk
Cooper's hawk
Sharp-shinned hawk
White-tailed kite
Kestrel
Merlin
Peregrine falcon
Osprey
Great horned owl
Barn owl
Screech owl
Northern saw-whet owl
Long-eared owl

mine) who sometimes came to watch me work. He would sit down, 100 feet off, and cock his head to one side, and watch me. I would take off my cap and bow to him, and then go about my work. Eventually when I looked up he would be gone, but I would see him again in a few days.

Animals as Pests of Crops

Contrary to my expectations, insect pests for the most part are not a problem on the farm. In spring, earwigs will chew on seedlings and flowers, but as the weather dries out in May they go into aestivation. Aphids regularly attack cabbage and broccoli, which are crops of which I grow only a small amount for the house. I have noticed that if I am going to grow 100 plants of broccoli, planted all together they will all get aphids, but if I plant five groups of twenty in five separate locations, a few of the plantings will escape infestation. By September cucumber beetles will start chewing on the petals of sunflowers; timely harvest evades this problem. Winter squash planted on a large scale (an acre or more) will surely attract squash bugs, so I grow only a few plants, widely spaced, which will escape. One year out of eight, olive fruit fly poses a minor problem. Careful management of pruning and irrigation is the first line of defense; an organic bait (spinosad) might be called for in a heavier infestation. Note that my management of insect pests consists primarily of avoiding susceptible crops; aside from a rare intervention for olives, use of insecticides is not necessary.

The animal pests most troublesome to me are vertebrates. Quail and sparrows will eat seedlings, especially in the spring, and some crops must be covered with a horticultural fabric to protect them. Wild turkeys are very keen on eating flower buds, especially anemones and ranunculus. Mockingbirds are fruit wreckers, and destroy a lot of figs. But it is rodents that are the most serious problem, and especially the western pocket gopher. The gophers dwell in burrows of their own construction, and eat the roots and stems of plants. They favor the junction of the root with the stem, their consumption of which kills the plant. They are particularly keen on olives and figs, and can kill a five-year-old tree with a stem diameter of five inches. The gopher, having felled the tree, does not stay to eat the whole thing, but instead moves on to the next tree, so that his feeding habit has an aspect of vandalism about it—kill an 80-pound tree to get four ounces of food.

As a protection against gophers, I plant some trees (olives and figs) in a fine-mesh wire basket that extends two feet into the ground and a foot above ground. Without this protection, I have had a planting of 98 fig trees entirely killed by gophers in less than a year.

Gophers are also destructive of flower bulbs. On several occasions I've had a single gopher eat more than $200 worth of tulip bulbs.

Ground squirrels are another difficult pest. They are seed eaters, and it is not so much their feeding that is a problem (although they will tear up a field of watermelons to get to the seeds) as their excavations. They dig their burrows deep and up to twenty feet long, and excavate many cubic feet, or even cubic yards of earth, which they pile up at the burrow entrances. This undermines the root system of trees, which they favor as a burrow location, and it impedes the operation of a tractor in mowing.

The largest rodents on the farm (other than the beavers in the creek) are black-tailed hares, which, surprisingly, do almost no damage except in the rare years when their population rises far above average. They will chew down a young olive tree, but other than that they just nibble a bit of grass here and there. And voles, which are most numerous, also do no notable damage.

Rats mostly do not bother the crops (though they will hollow out a ripe pomegranate on the tree), but stick to the buildings, as is their habit. They can be very destructive. Twice a rat has got under the dashboard of my van and chewed up the wires; I've had to have it towed to a garage for expensive repairs. Rats chewing on electrical wires set fire to a trailer and barn nearby. They get into the henhouse at night where they steal feed and disturb the hens. And their scurrying about in the walls and attic of the house keeps one awake at night.

I prefer not to kill animals, but I set mechanical and electric traps for rats. I trap squirrels live in little cages that I've made, with one-way doors, and which I bait with nuts and seeds. I release the squirrels at a refuge a few miles away. Problem gophers I trap using little wire traps that I set in their burrows. The traps have a spring which squeezes the gopher's thorax so that when he exhales he cannot then inhale, and he suffocates. There was a time when I put a lot of effort into trapping gophers, but it was like taking water out of the ocean with a teacup. Now, if I have a particular problem gopher in a sensitive area I'll set a trap, but otherwise I ignore them. The best use of the farmer's time for managing rodents is the time spent constructing nest boxes for barn owls. These are quickly occupied, and the owls are more effective than the farmer in dispatching the troublesome rodents.

The economic cost of rodent pests, almost all of which is attributable to gophers and squirrels, is significant. I may lose 6 percent of my gross income to rodents, which doesn't sound so bad, but that translates to 25 percent of my net income. And when one is already financially marginal, a 25 percent loss is painful. There have been several years, about one in ten, in which the rodent populations rapidly rose to very high levels. This includes all the species—voles, gophers, squirrels, rabbits, and hares—suggesting that a dearth of predators might be a factor. In those years I may lose up to 60 percent of my annual crops (more for some species), although the perennials are not affected.

Comparison with Conventional Farms

Pest problems on conventional farms growing row crops and field crops are very different from the pest problems on my farm. Consider a 600-acre field of wheat farmed conventionally. The crop is harvested in June, and the straw baled soon thereafter. The ground is worked up and listed into beds that will be planted to sunflowers the following April. For nine months this is bare ground. For a vertebrate, there is nothing to eat, no water available, and no place to hide. If there were rodents in the wheat field, they either migrate or die. With fall rains, weeds and volunteer wheat germinate—these are killed off with herbicides. Finally, in April, the sunflowers are planted.

There is no way that vertebrates can migrate and reproduce fast enough to become significant pests of this field before the crop is harvested—a matter of months. But insects can colonize it very rapidly, especially in the absence of predators that might have suppressed the infestation in a diversified farm. So for the conventional farmers, insects are the chief pest problem, and the farmers deal with this by heavy application of pesticides. Rachel Carson's 1962 book, *Silent Spring*, referred to this situation, and created a widespread impression in the public that insects were the principal pest problem of agriculture. However, on a biologically diversified farm like mine, insect pests are a very minor problem compared to vertebrate pests. Not only are insect pests not much of a problem, but a great variety and density of native pollinators (Hymenoptera, Syrphidae) established on the farm ensure successful pollination of crops without the need to rent honey bees.

Biodiversity

Analysis of biodiversity is extremely sensitive to scale. To a beetle trudging along on his little legs, six tomato plants in a row constitute a monoculture. And to a butterfly on a windy day, 40-acre plots are a polyculture. For the present purpose, taking 20 acres as the unit of analysis, my farm has about 100 times more biodiversity of plants and animals than most of the farms in the region. Partly this is the influence of the creek, and partly it reflects my scheme of planting, not only crops but also a great number of ornamental plants that I find to be interesting in their own right despite their lack of commercial value.

The conventional farms operate on the principle of "total control through simplification." A single crop of genetically identical individuals is grown, and one can in theory determine exactly the requirements of nutrition, irrigation, pest management, and disease control, that result in maximum profit, which usually equates to maximum yield. Whereas I follow the principles that diversity creates stability, that complexity of the system cannot be entirely known, and that less intervention is better. I am not suggesting that my system is superior. The conventional tomato growers in my area produce a harvest of fifty tons per acre that they can sell profitably at 3.6 cents per pound, and the tens of thousands of acres of tomatoes farmed in this way put a can of tomatoes on the shelf in the store for 89 cents, not twelve dollars. There would be serious consequences to suddenly abandoning such a finely tuned system. Nonetheless, agriculture in the valley as a whole seems to be heavily unbalanced in favor of conventional farms over biologically diversified farms.

Chapter 5

Care of the Soil

Soil is not dirt. If it is healthy, it is a complex ecosystem with an intricate three-dimensional architecture occupied by an abundance of organisms: bacteria, fungi, actinomycetes, algae, invertebrates, reptiles, mammals, and the roots of vascular plants. Caring for the soil is the farmer's number one task; if the soil is healthy, the crops will look after themselves.

A useful measure of soil health is soil respiration, that is, the collective breathing of all the soil organisms, including the roots of plants. If the soil is dry, or cold, or too alkaline, or too acidic, or deficient in oxygen, or too compacted, or deficient in critical nutrients, or deficient in organic matter, or contaminated with toxic substances, then respiration will be diminished. Although there are laboratory methods for measuring soil respiration, one can get a good sense of it simply by examining the soil. It should be moist, warm, in good tilth, of good odor, with an abundance of earthworms and arthropods and the roots of plants.

Origins of the Soil

According to the well-driller's log, the layer of loam soil on my farm is 314 feet thick, at least at the spot where he was drilling a well. This is typical of the region, and is part of the motivation for the shift from row crops and field crops to tree crops. With soil that deep, it seems wasteful to use only the top six inches.

The soils are of alluvial origin, that is, they were deposited by rivers. The Sacramento Valley is surrounded by mountains—the coast range to the west, Mount Shasta to the north, and the Sierra Nevada to the east—and rivers arising in these mountains carried sediment into the valley to form the mineral basis of its soils. We are taught in school that erosion is bad, and this is true in the place where erosion is occurring. But erosion is two-ended, and the eroded materials will eventually be deposited at some other place, which thereby benefits from erosion. And so the rich soils of Mesopotamia were the result of erosion in the highlands of Armenia, and the life-giving soils of the Nile delta were the result of erosion in west and central Africa. Similarly, the soils of my farm are the result of erosion in the coast range of mountains.

The ability of moving water to carry sediment depends on the velocity of the water and on the steepness of the slope down which the water runs. As water enters the valley, it slows down. Large particles are dropped first, then medium-sized particles, and finally the tiniest particles of clay. And so we may find sandy and gravelly soils close to the mountains, and clay soils in the center of the valley. The clay soils in the valley basins are in regions of still waters, and these have become rice-growing districts. My farm is at middle distance from the mountains, and the soils are a silty clay loam.

Only about a third of the earth's land surface is suitable for growing crops, and there is a certain logic to moving needed materials from the other two-thirds into the arable regions. Erosion is one way to do this, but intentional interventions are appropriate as well. Minerals (phosphate rock, gypsum) can be mined, as well as organic matter (peat, for example) to be used as soil amendments on arable fields.

Soil Texture and Soil Structure

The term "soil texture" refers to the proportions of mineral particles of various sizes in the soil—sand, silt, and clay. This is not easily altered, so whatever soil texture you have when you buy your farm is the texture you will always have. An important aspect of soil texture is the ability of the soil to hold water and to accommodate air. Heavy soils (clay) are usually highly water-retentive, but often are poorly oxygenated, whereas sandy soils are the opposite. The water retentiveness of a soil is a critical determinant of irrigation practices.

The soil on my farm is a silty clay loam that has some sand, mostly silt, and a bit of clay. At one time I considered putting up a building with rammed earth walls, using soil from the site. I made various trial blocks of rammed earth, including some strengthened with Portland cement, but they were inadequately strong. Successful rammed earth requires a higher percentage of clay than what my soils have.

"Soil structure" refers to the three-dimensional arrangement of the soil components, and is very much under the farmer's control. The distribution of organic materials of various sizes, the aggregation of soil particles into crumbs, the presence of open channels created by invertebrates or by the death and decay of plant roots, and the percentages of water and air in the soil are aspects of soil structure. Tillage, irrigation, cover cropping, amending the soil with minerals or organic materials, mulching, and avoiding compaction are tools that the farmer has in caring for soil structure.

When school children visit my farm, I make a point of digging up some soil that has not been tilled in twenty years or more and passing around chunks so that the students can see the structure of the soil, which is complex and interesting, riddled with little open channels, and fragments of roots, and the dens of terrestrial insects. If I till an acre of ground in the autumn, plant an overwintering crop, and harvest it the following spring after which the soil is left undisturbed, I can then dig up the soil a year after the tillage to see how the structure has developed. The finding is that there is hardly any structure to the soil; one year is not nearly enough to begin the development of a soil ecosystem. It is a matter of many years for soil structure to develop, and a single-year fallow, for example, while helpful to soil fertility and organic matter, is of little use in developing a mature soil structure.

Tillage

When most people think of farming, they think of tillage—the farmer out on his tractor, plowing or cultivating. And for modern industrial farming of some crops (such as grains), tillage constitutes the majority of the farmer's activity. Many ingenious implements have been invented for chopping up and rearranging the soil, and such devices make up the major portion of the farmer's equipment. The goals of tillage are to loosen the soil, disrupt weeds, and incorporate amendments in order to enhance growth of the crop. However, tillage has significant negative consequences for the soil, including disruption of the soil ecosystem and accelerated loss of organic matter. On balance, tillage is almost always a net negative for soil health, and is to be avoided when possible.

In the absence of tillage by humans and their machines, the soil is not static. On my farm, burrowing rodents are highly active at moving soil about; I have calculated that if the activity of rodents were spread out evenly, they would till the entire top foot of soil in about 23 years. Earthworms and arthropods are also active movers of soil. The action of water and wind can move soil about, although flatness of the land greatly inhibits water-based erosion.

Once I was dismantling a house, built in 1867, for salvage, and as I reached the crawl space under the floor I found that the level of the ground was about a foot lower than the surrounding land. I do not believe this was excavated at the time of construction. Rather, wind-blown dust hitting the sides of the building had fallen to the ground, and over a century and a half had raised the level of the earth by a foot. The dust was made airborne by plowing dry land on windy days.

On my farm I have been only partially successful in eliminating tillage. The apricot orchard is mowed but not tilled. In the olive grove, I am forced to till once every few years to break up soil compaction. Olive harvest takes place during the wet season, with the result that a lot of activity in the grove, including tractors pulling heavily laden trailers, is on wet ground, resulting in compaction of the soil in the alleys. Tillage is limited to a strip six feet wide up the alleys; the tree rows are heavily mulched, and are not tilled.

For my annual crops I have not found a good alternative to tillage. Cover crops grown over the winter and irrigated by winter rains are mowed in the spring with a flail mower. The ground is then worked with a chisel plow, followed by a disk. When suitable soil moisture is achieved, amendments (compost, gypsum) are spread, and the top few inches are rototilled into a fine tilth to serve as a seedbed. Several times during the season the crops may be cultivated using a sled fitted with knives and sweeps, or the small cultivating tractor equipped with mid-body toolbars. When the crop is finished, it is mowed and then the ground disked and harrowed. The cover crop seed for the following winter is broadcast and covered by a harrow, and the land is firmed with a cultipacker.

The effects of tillage are best understood by considering separately the short-term effects and the long-term effects. In the short term tillage incorporates surface organic matter into the body of the soil, and by breaking up organic materials exposes new surfaces to bacterial action. The soil structure is rearranged, with increase in oxygen that stimulates breakdown of organic matter, accelerating the release of nutrients. In addition, the more open structure facilitates growth of the crop's roots. These factors are the rationale for tillage prior to planting annual crops—the more open structure and enhanced soil respiration stimulate crop growth. There are also short-term negative consequences to tillage: death of soil organisms, stimulation of weed growth, and increased risk of erosion by wind or water.

The longer-term effects of tillage are less benign. The temporary spike in soil respiration that stimulates the crop ends up decreasing net soil organic matter. Historic comparisons of virgin soils to adjacent cultivated soils have shown that cultivating always leads to a loss, sometimes quite severe, of soil organic matter. And with only a few exceptions, tillage is an accelerant of soil erosion.

I have made several attempts to decrease tillage for my annual crops. One was to use strip tillage for melons and cucumbers, which involves mowing the cover crop and then tilling a narrow strip—six or eight inches—every six feet or so where the crop will be planted. The idea is sound, but I encountered two problems. The first is that I don't have appropriate equipment for creating a suitable seedbed in the tilled strip. This is not insoluble—I have been on the lookout for a used tractor-mounted rototiller from which I could remove all of the tines except for one gang in the middle. The other problem is that the grasses (oats, rye, millet) in the cover crop mix continue growing, sometimes quite vigorously, after they have been mowed. I would need to shift to a grass-free cover crop for strip tillage. These are simple problems, but I have not overcome the inertia that impedes their solution.

The other approach to decreasing tillage is to go to a system of permanent raised beds. The raised beds permit furrow irrigation (not relevant for me) as well as drainage in heavy rains. Traffic of tractors and humans is confined to the furrows, so that compaction of the beds is mostly eliminated. Accumulated compaction in the furrows, on the other hand, suppresses the growth of weeds. Another advantage of permanent raised beds is that expensive soil amendments such as compost and mulch are not wasted in the furrows, as would occur if the compost were spread evenly over the field before the beds were laid out.

Where I have long-lived perennial crops, such as peonies, a system of permanent raised beds has worked well. But for annual crops my determination is not strong enough to persist with permanent raised beds, and sooner or later I plow the whole thing down in order to densely sow a green manure crop and start over. Again, lack of suitable equipment plays a part. A very narrow grain drill would let me sow my cover crop to the tops of the beds, and a narrower flail mower than the one I use in the orchard would allow me to mow just one bed at a time.

There are several reasons why the failure of no-till and strip-till practices on my farm is not highly important to me. For one, much of the literature demonstrating the superiority of no-till/strip-till derives from the Corn Belt, and was based on comparing no-till to deep plowing with a moldboard plow, a practice that we have long known to be destructive of the soil (see Faulkner's *Plowman's Folly*, 1943). Shallow tillage with chisels and disks is far more benign than deep plowing, and is only slightly detrimental compared to no-till. Second, the soil must be tilled at least somewhat (even use of a seed drill is a form of light tillage) in order to plant cover crops in the autumn. And third, in this climate, the vital activity of the soil declines to near dormancy during the dry season of summer. Earthworms ball themselves up in little chambers a foot or more below the soil surface where they await the return of moisture, and other microfauna and microflora also aestivate. Under these conditions, light tillage of the soil surface is a reasonable practice with few negative consequences. Indeed, it can help to conserve deep moisture.

Soil Compaction

Health of the soil requires a reticulum of pore spaces that facilitate soil respiration, allow rapid movement of water into the soil, and provide channels down which the roots of plants may grow. Compaction of the soil results in collapse of these passageways, starving the soil ecosystem of oxygen and impeding the passage of water. Sandy and gravelly soils are fairly resistant to compaction, but loams and clays are susceptible, especially when they are moist.

Farm equipment is the chief culprit in soil compaction, and there is a paradox connected with this. The heavier a tractor is, the better its traction, and so we add iron weights and fill the tires with water to increase the tractor's mass, which in turn worsens compaction caused by the tractor, therefore requiring additional tillage. Damage to the soil from compaction by machinery can be localized if all of the machines have the same tread width and are driven on the same paths. This is relatively straightforward with modern, GPS-guided equipment. Permanent raised beds, mentioned above, are another way of confining compaction to specific zones.

There is another sort of compaction that occurs on a smaller scale, and that is local compaction at the soil surface caused by raindrops. Raindrops strike the soil with considerable force, with two results. The first is simple compaction. The second is that the soil in the top few millimeters momentarily enters an aqueous suspension, which results in a rearrangement of the soil particles, with the lighter particles (clay and silt) moving to the top, and with the heavier particles underneath.

When the soil then dries out, a crust of dried clay and silt is formed, which limits movement of air and water. This process occurs on bare earth. A simple layer of organic matter on the surface, even a single leaf, diverts the water's force and inhibits crust formation.

The most problematic time for managing soil compaction is in the spring. The greenhouse is full of overgrown seedlings becoming rootbound in their trays; they desperately need to be planted, but the ground is too wet. We get a couple of dry days and I think, "Maybe I can get away with this," and so I take a tractor out into the field hoping to prep a few beds, but in fact the ground is still too wet, and the result is compaction that will take a year or more to remedy.

The most severe compaction I've seen on my soils was caused neither by machinery nor by rain, but by a flock of escaped sheep trotting through the orchard when the ground was wet.

Soil Fertility

The prosperous growth of plants requires that an array of mineral elements be available in the soil. A deficiency or imbalance of these will be detrimental to plant growth. The common advice the farmer gets is to send soil samples to a laboratory for analysis, which will provide the information needed to choose a course of remediation. This is true, but there are other means of reaching the same conclusions.

One is to study your crops. Classic mineral deficiencies or toxicities have well-defined manifestations in the crop, and if you are knowledgeable you can use these to diagnose fertility problems. Many soil fertility problems are regional, and so your neighbors or your farm advisor can suggest likely scenarios without any lab work. If, for example, all of your neighbors have issues with zinc deficiency, you can be pretty sure that will apply to you as well.

Once you have a hunch about a mineral imbalance that requires correction, you can get advice from a laboratory, or you can embark on a program of experimentation. For example, in the case of suspected zinc deficiency, you might purchase some zinc sulfate fertilizer and try three different rates of application on a small area of the crop in order to gauge the response.

In my region, an excess of magnesium and a shortage of calcium is a common problem, so that I was aware of it before I began growing any crops. One of the most dramatic manifestations of this imbalance is seen in persimmon trees, where the leaves fail to turn dark green, but instead remain yellowish for a long time, eventually becoming a pale green. So the message from the persimmon trees with their pale leaves, together with the regional history of magnesium/calcium imbalance, made it clear that I needed to be adding calcium. I sent soil samples to a laboratory, and they came back with a reading of 889 ppm for magnesium, which is very high, and 1,162 ppm for calcium, which is low. The remedial action is to add gypsum (calcium sulfate), which is an inexpensive powdered rock mined in various western states. It can be applied at very heavy rates without causing toxicity, and so I spread 50 pounds of gypsum around each persimmon tree, repeating this every second or third year. A year after the first application; the persimmon leaves were dark green. Incidentally, the laboratory report included a recommendation of applying two tons of gypsum per acre, which is typical in this region.

This example demonstrates the mutual reinforcement of knowledge of the regional situation, the classic symptoms in the crop, the laboratory analysis, and the experimental remediation.

The laboratory report includes many other measurements, most of which are unsurprising, except for levels of boron, which were very low at only 0.6 ppm. Boron is an element for which the difference between too much and too little is very narrow. Since I was not observing any sign of boron deficiency in my crops, I chose to take no action because of the risk of unintentionally over-applying it.

It is important to note that the presence of an element in the soil is not the same as its availability to the crop. There may be abundant phosphorus in the soil, but if nearly all of it is locked up as aluminum phosphate, which is unavailable to the plants, then the crop may still show a phosphorus deficiency. The key to availability is to be found in the life of the soil—specifically, microorganisms metabolizing organic matter. All of the conditions previously mentioned that lead to high soil respiration—oxygen, moisture, warmth, organic matter, developed soil structure, and abundance of organisms—are also the requisites of soil fertility.

From this perspective, the best means to improve soil fertility is to add organic matter. This takes two forms. The first is growing green manure crops—in my situation in the winter rainy season—and then plowing them down in the spring. Generally I grow a mixture of legumes and grasses. The legumes fix nitrogen, making it available to plants, but the grasses provide considerably more organic matter. The most recent laboratory report from the tilled soils where I grow annual crops shows adequate nitrogen but low organic matter (2.9 percent), which suggests changing the composition of the cover crops in favor of more grasses and fewer legumes.

The second method of increasing organic matter is to truck it in. I am lucky in that within 40 miles of my farm there are four commercial compost facilities from which I can buy truckloads (40 cubic yards) of organic compost at a modest price. The compost usually has a nitrogen content of about 3 percent, although much of this is ammonia, which is likely to outgas before the compost is incorporated in the soil. In addition to supplying nutrients and a substrate for the growth of microorganisms, the compost improves soil tilth, making it easier to transplant seedlings or to plant bulbs.

Adding organic matter to the soil is a job that is never completed. The high rate of soil respiration in a healthy soil reflects oxidation of organic matter by soil organisms. Under optimal circumstances of abundant oxygen, moisture, and warmth, soil organic matter burns up quite rapidly, and so requires continuous replacement.

This conception of soil fertility is not shared by most of the conventional farmers in my area. They consider the soil to be dirt, and they despoil its structure and its vitality with excessive tillage, compaction, application of pesticides, long bare-ground fallow periods, and application of corrosive fertilizers. Having destroyed the microorganism-mediated pathways of soil fertility, they substitute heavy applications of chemical fertilizers. The least expensive form of nitrogen locally is anhydrous ammonia, which is a gas at atmospheric pressure, but which is usually handled as a liquid under high pressure. It can be shanked into the soil, where it is extremely toxic to soil organisms, or it can be injected into irrigation water. Soils farmed in this way for a number of years become almost entirely depleted of soil organisms and will have very low levels of soil organic matter.

The Net Annual Change in Soil Fertility

Soil nutrients can be lost from the farm by soil erosion (especially true of phosphorus), which is not an issue on my farm; by leaching out of the root zone into groundwater; and by harvesting crops that are sold off the farm. Because this is a region of scant rainfall, and because I do not over-irrigate, and because many of my crops are deep-rooted (e.g., olives with roots at 20 feet), I expect that losses to leaching are minor. This leaves harvesting and sale of crops as the principal mechanism by which the soil fertility of the farm might be diminished.

About half of the cultivated land is planted to olive trees, and the olives are processed on-site to extract olive oil. Chemically, olive oil is just carbon and hydrogen, with almost no minerals in it; all of the nutrients—nitrogen, potassium, phosphorus, calcium, etc.—are in the waste material from the mill ('pomace') which is spread back into the olive grove as a nutrient-rich top-dressing. So in selling olive oil as a major product of the farm, I am removing hardly any nutrients from the soil. In fact, because I also mill olives for other growers who bring me their fruit, I end up with a net increase in nutrients due to olive oil production. When I was doing custom milling on a larger scale I acquired an extra 20 to 30 tons of pomace each year, which was spread on the open ground as well as in the orchards. I have subsequently scaled back this activity, but it continues to supply a net increase in soil nutrients.

In the Mediterranean region, olive pomace is considered to be a highly problematic form of biologically hazardous waste. The reason that is not true for me has to do with scale. A big olive mill processing the fruit from 30,000 acres ends up with a tremendous amount of pomace that cannot easily be returned to its place of origin, and this accumulation presents difficult problems of safe disposal. Because I am working at a small, artisanal scale with respect to both volume of pomace and distance from the grove, the pomace is easily spread at a horticulturally benign rate. As in pharmacology, the dose makes the poison.

I plant about half a ton of flower bulbs each year, and here too the nutrients delivered to the farm within the bulbs probably just about balance the nutrients lost by the sale of the flowers. Tulip bulbs can be grown to maturity without soil, but just in water, indicating that the nutrient content of the bulbs was sufficient for production of the crop.

Nutrient loss from sales of crops then must be attributed to the crops other than olives and bulbs: berries, apricots, persimmons, melons, cucumbers, citrus fruit, and flowers grown from seed. I do not have a means to calculate the nutrients lost in this way, but I can estimate the net weight of products sold as on the order of 10 to 20 tons, wet weight. Each year I purchase and have delivered 20 tons (dry weight) of commercially made organic compost (urban green waste plus restaurant waste), which likely has at least as great a nutrient density as the crops sold. So, simply based on weight of crops sold versus weight of compost purchased, I am probably net positive for soil nutrients. In addition, I acquire varying amounts (2 to 10 tons per year) of wood chips for mulch, which, though low in nutrients, represent a slight increase for the farm. Roughly half a ton of gypsum is added each year, increasing net density of calcium. And legume cover crops grown each year help to maintain nitrogen levels.

The upshot is that soil fertility as measured by nutrient content of the soil and soil water is probably increasing year by year, even without the addition of any chemical fertilizers. I should point out that urban green waste (grass clippings, leaves, etc.) derives from plants that for the most part have been heavily fertilized with synthetic chemical fertilizers. When I buy organic compost made from urban green waste, I am simply getting synthetic chemical fertilizers one step removed. If you have an appreciation of geochemical cycles, then there is nothing surprising or disturbing about this, despite the highly charged debates about chemical versus organic sources of nutrients.

Mulch

In the natural soils of forests or grasslands the naked earth is seldom seen. The surface of the soil is covered by a layer, sometimes quite thick, of dead and decaying plant materials. In agricultural situations we can simulate this by laying down an organic mulch.

The mulch keeps the soil cool in hot weather and protects it from freezing in cold weather. It obstructs the germination of weed seeds. It protects against compaction by raindrops, and retards evaporative losses. And the interface between the mulch and the soil surface is biologically an extremely active zone for both microorganisms and invertebrates.

Mostly I use wood chips for mulch, which I buy from the local contractors who prune trees for the power line right of way. I also use rice straw, which is the cleanest and least expensive type of straw available locally. These products have a high carbon/nitrogen ratio and so are slow to decompose, but because they are on the surface of the soil they do not cause the short-term decrease in available nitrogen that they would if they were incorporated.

In my orchards, the branches pruned from the trees are shredded into mulch. Years ago, prunings would be gathered into a large pile and burned. But because this was detrimental to air quality, the practice was outlawed. In response, an industry of custom shredding grew up. When my pruning is finished, I call a local contractor who brings in a large machine that runs on tracks and is powered by a 400 horsepower diesel engine. It drives through the orchard at a pretty good clip, sucking up the branches and leaving behind an even mulch of chips about the size of matchsticks.

The word "mulch" is inappropriately used for layers of plastic sometimes rolled out on the surface of the ground to suppress weeds and to increase soil temperature in cold weather. But the plastic mulch suppresses soil respiration by blocking the movement of oxygen and water into and out of the soil. And as a petroleum product it entails additional energy and ecological costs.

Chapter 6

Water

The region of my farm has a classic Mediterranean climate, with nearly all of the rainfall occurring in the five-month period from November through March. The summer months lack significant rain. For a farmer, there are several advantages to this climate. The cloudless skies of the summer months ensure abundant intense light. Dry weather during the growing season is inimical to plant diseases, most of which require high humidity to spread aggressively. Having undertaken the labor and expense of irrigation, the farmer is compensated by complete control over the timing, location, and volume of water applied to his crops. Finally, farm logistics are not hampered by untimely rains. This is particularly important for the major row crop in my district—processing tomatoes. Contracts with the canneries specify harvest dates a year in advance so that the cannery can operate at full capacity without being short of fruit one day and overwhelmed the next. And with the fields and farm roads dried out enough to support heavy equipment, the harvest can proceed exactly on schedule.

Balancing its advantages, the Mediterranean climate also has a particular drawback. The requisites for robust plant growth—light, heat, and water—are out of synchrony, so that water is abundant when it is cold and dark, and water is insufficient when light and warmth are abundant. This is evident in the accompanying

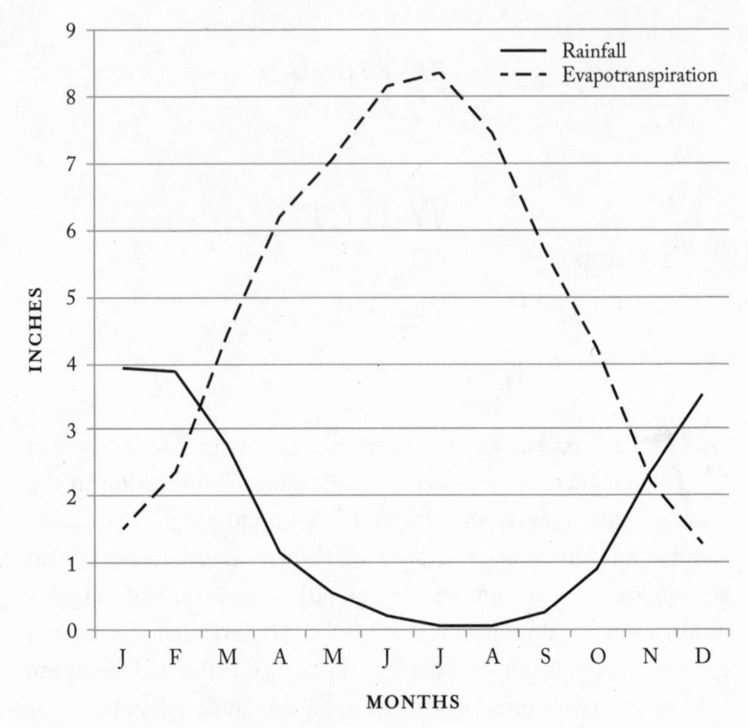

Rainfall and evapotranspiration by month. Based on rainfall data from UC Davis; Eto data from CIMIS; graph by Emre Kara.

graph, showing rainfall as well as evapotranspiration (Eto), which can be thought of as a measure for the crop's demand for water. Certainly the native oak savanna in this region was a water-limited ecosystem whose annual productivity was less than what the levels of light and warmth could have supported. And the increase in productivity of agriculture is entirely dependent on irrigation.

The water-retentive features of different kinds of soils mentioned in the previous chapter are pertinent here. Sandy soils have little capacity to store water, while the heavier silty loam and clay loam soils can sequester a large portion of the winter rains and slowly release the stored moisture to deep-rooted plants as the dry season develops.

Rainfall Averages

Unlike the stable, oceanic climate of an island where rainfall amount is pretty much the same every year, the continental climate of the Sacramento Valley has a high variance in annual rainfall. Annual totals in the last century vary from 7 to 38 inches of rain. In looking over historic rainfall data for this region, I noticed an odd thing. For the 142 years from 1872 to 2014, the average rainfall was 17.8 inches, but the median was only 16.1 inches. Sixty-one percent of the years had below-average rainfall. Another set of data, local monthly rainfall for 132 months from 2004 to 2014, showed a similar situation: 71 percent of the months had rainfall below average for that month, and only 29 percent were above average. What this shows is that rainfall does not follow a Gaussian distribution, but is skewed by rare episodes of very high rainfall.

One consequence of this is that if you are establishing your expectations for rainfall in the coming season based on historical data, you should expect the median rainfall (16.1 inches) rather than the average rainfall (17.8 inches). The average reflects the statistical influence of rare, super-wet years.

We might question the ecological consequences of unusually wet years. For one thing, high volume of runoff clears debris out of the creeks and other channels—we saw this in the super-wet year of 1998. The high runoff also fills lakes and reservoirs, the water of which may then be carefully distributed over a number of years. And it seems plausible that in exceptionally wet years the soil profile can be charged to field capacity at a much greater depth than occurs in years of average rainfall. This may result in some loss of soil minerals to leaching, but it also would be highly beneficial to deep-rooted perennial crops.

Sources of Water on the Farm

The median rainfall of 16.1 inches is sufficient to grow a number of tree crops, provided that they are widely spaced so that their roots can draw moisture from a large volume of soil. Much of the production of almonds and olives in Spain, the Levant, and North Africa is on unirrigated land with rainfall of this magnitude. However, the yields are extremely low compared to the same crops grown in California with irrigation. Dryland hay crops can also be produced without irrigation, although the yields and value are sometimes quite low, and some years rainfall will be insufficient. Historically, barley was sometimes grown following two years of cultivated fallow that allowed the moisture level in the soil to build up.

The advantages of rainfall are that the water, originating in the northern Pacific Ocean, is of good quality, it is free, and it falls without prejudice over all parts of the farm. Each winter I collect from the roof of our house 1,000 gallons of rainwater, which we store in a steel tank. This is used for drinking, and also serves as an emergency water supply if there is a problem with the well.

A government-funded irrigation district provides very inexpensive (i.e., subsidized) surface water to local farmers from a reservoir in the coastal mountains. This supplies many farms near us; however, our farm is not in the irrigation district, and so our household water and our water for irrigation are pumped from groundwater.

There are two wells on the farm, one at a depth of 200 feet, and one at 340 feet. The deeper well has a 1.5 horsepower electric pump; the shallower well had an antique water-pumping windmill that was ineffective because of its inefficient design and the lack of wind. That well now has a direct-current pump powered by five solar panels (1,400 watts); its operation is off the grid. Each of the wells produces between 20 and 30 gallons per minute from a static water level of about 60 feet below the surface. Both of the wells are considered domestic wells because of the small diameter of the casings (6 inches) and the low power of the pumps. Agricultural wells in this area are much larger, and usually deeper (to 1,000 feet or more), and with powerful pumps that may put out 1,000 to 2,000 gallons per minute.

Quality of Water

The soil from which my water is obtained is a three-dimensional mosaic of gravel, sand, silt, and clay. The goal of the driller is to find water-bearing deposits of gravel and sand where the screened sections of the well casing are placed. Wells deeper than about 300 feet enter into a different geologic zone, with ancient water that is rather different in character from the water of shallower wells.

The quality of water from my wells is typical for shallow wells on the west side of the valley. The pH ranges between 8.0 and 8.3. Levels of magnesium (45 ppm) and calcium (37 ppm) are very high—mostly these are the carbonates—and hardness (278 ppm) is almost double the level that would define "very hard" water. Any solid surface hit by a sprinkler will turn white from the dissolved salts, and plumbing fixtures rapidly build up hard white deposits. Locals joke that you can set a pan of water out in the sun to dry, and make bricks that way. The problematic high levels of magnesium in my soils undoubtedly have been worsened by a century of irrigation with high-magnesium well water. Although nitrate pollution is a problem with many wells in the valley, our water is okay on this score, with a low concentration of 12 parts per million. The wells also pump a small amount of sand and silt, which is problematic for drip irrigation emitters.

Geologically ancient water from deeper wells in the region (ag wells and urban wells) has a somewhat different quality. It has much lower levels of magnesium and calcium and is less hard, but it has high levels of sodium that can be damaging to crops. It also tends to be high in arsenic, making it less useful for livestock and humans.

The Irrigation Infrastructure

Water from my wells is distributed around the farm through PVC pipes of 1.5 to 2 inches in diameter, buried at a depth of about 20 inches. A map of the system would appear illogical, for it was not built all at once, but was added onto bit by bit over a period of years. Each addition was connected by the shortest path to the existing pipeline, which is logical, but the end result is needlessly long and contorted pathways for the water to reach its destination.

At various places, riser pipes are brought to the surface with a valve where drip or sprinkler irrigation can be attached. The orchards have drip systems built with heavy-gauge tubing and button emitters designed to have a lifespan of at least twenty years. Annuals and other herbaceous crops are irrigated using lightweight drip tape, which seldom lasts more than one year. The cause of the short lifespan is not so much the inadequacy of the material but the accumulation of damage due to rodents seeking water. By the time a 200-foot length of drip tape has six couplings in it (representing six repairs) it is more efficient and more profitable to replace it with a new line. I generally replace about 8,000 feet of drip tape each year. The retired tape is taken to a recycling facility.

The pressure of the water in the main lines varies from 40 to 60 pounds per square inch (psi). The drip systems operate at a pressure of only 10 to 20 psi. At higher pressure, the drip tape simply blows up. So I use pressure reducers at each valve to bring the pressure down to about 15 psi. I could use a single large pressure regulator at the wellhead, but I prefer the decentralized system that lets me tap high-pressure water at any valve if I happen to need it.

The conventional wisdom (and manufacturer's recommendation) is that you lay a line of drip tape down each row of plants. But I prefer one line of tape for every six or eight rows of plants. When irrigating, I leave it on a row for two or three hours, and then move it to the next row. This has the advantage of using less tape, but the greater advantage is that the tape is out of your way when you're cultivating or weeding. Drip tape and cultivation do not coexist well, and I prefer to keep the tape out of the way except during irrigation.

Operation of the system is not as trouble-free as you might expect. Every time a section is turned on, I have to walk the full length of it inspecting the lines and the emitters for damage.

One type of damage consists of leaks caused by animals. But it is also common that some emitters are clogged. What happens is that minute amounts of sand or silt get into a valve such that it cannot be completely turned off, but continues to allow a small amount of water to pass. This is true of both ball valves and gate valves—plastic, brass, or stainless steel. The small volume of water in the system results in a continuous wet drop at each emitter. In hot weather, evaporation from this drop can lead to accumulation of salts (magnesium carbonate) that seal off the emitter so that next time you open the valve, no water comes out of the emitter. The other thing that happens, quite surprisingly, is that spores of mosses germinate in that drop of water and start to grow, and the emitter becomes clogged with a tiny moss plant. This happens in the olive grove, where mosses grow near the lower trunks of the trees.

The maintenance requirements of the drip irrigation have led me to reconsider the advantages of furrow and flood irrigation. Long use of drip irrigation can lead to accumulation of salts in the wetted zone. The tree crops certainly appear to be less stressed with flood irrigation, which reaches all of their roots rather than just the few spots reached with drip. And when there is a series of drought years through which the whole soil profile gradually dries out to the wilting point, the value of flooding is obvious. So now, in drought circumstances, I use a combination of drip irrigation and furrow irrigation (five shallow furrows one foot apart) for the olive trees. I plan to do that for the apricots as well.

Operation of the Irrigation Program

I am a frugal irrigator. Each year I pump about 900,000 gallons of irrigation water. If this were spread evenly over the cultivated portions of the farm it would achieve a depth of about 2.5 inches. Adding the 16 inches of rainfall gives a total of only 18.5 inches of water applied to the land, an unusually low level for this region. By comparison walnuts, which are grown on three sides of my farm, require 44 inches of water for maximum yield, so that in addition to the 16 inches of rainfall, the farmer must apply 28 inches depth of irrigation—more than ten times what I am using. In order to provide this much water, the big agricultural pumps operate day and night.

I do not apply the irrigation water evenly over the land; application is strategic both in space and time. Much of the open ground is dedicated to winter-growing crops, which in most years are matured without any irrigation. And for summer annual crops, irrigation is applied selectively. Two rows of sunflowers with a drip line between them constitute a bed; the next bed is five feet away, with no irrigation between the beds. Watermelons are impressive scavengers of water; a single deep irrigation at planting time is sufficient to mature a crop.

Of the tree and perennial crops, citrus, persimmons, and blackberries require a sustained high level of irrigation. Apricots, which can be dry-farmed, are not irrigated until after harvest in June, and then receive about five irrigations through a drip system. Olives can use about 36 inches of water, so that the deficit irrigation practice recommended by the university—70 percent—would require 25 inches. I am providing a total of about 20 inches—56 percent of maximum useful irrigation.

There are a number of reasons to skimp on irrigation. One is simply a dislike of wastefulness and the desire to conserve a scarce resource (while reducing the economic cost of irrigation). It is also true that more water means more weeds. But the most compelling reason is that drought-stressed fruit has the best flavor. Growers of wine grapes were among the first to understand this, and the best wines are made from vines suffering considerable water deficit. The effect is equally true with stone fruits. Compare apricots from a heavily irrigated orchard—big, beautiful, and tasteless—to fruit from drought-stressed trees—small, dense, and highly flavored. Olives also respond to water deficit with increased flavor and levels of polyphenols. Olive oil from abundantly irrigated groves tends to be bland, with a short shelf life. It is possible to overdo this; olive oil from severely stressed dry-farmed trees may be extremely bitter.

The relationship between drought stress and increased flavor does not apply to all crops, and in some cases heavy irrigation for maximum yield seems appropriate (assuming that water is available). Hay would be an example. Nut crops (walnuts, almonds, pistachios, chestnuts) appear not to show any increase in quality in response to water deficit, and may be grown for maximum yield. The same is true of citrus fruit and blackberries, for which drought causes a decline in quality.

CHAPTER 7

Farm Infrastructure

When I purchased my farm, it had no buildings, no well, and no electricity, and the access road was a seasonal dirt farm road. A 12-inch-diameter pipeline in one corner was meant to bring irrigation water from a distant well; however, that well, a very old one, failed during the first year with a collapsed casing and was abandoned.

A farm needs more than just good soil to carry out its mission. The requirements include all-weather roads, buildings for secure storage of supplies and equipment; packing facilities for preparing crops for market; greenhouses for starting seedlings; cool storage; electricity for light, power, and pumping water; wells to provide a source of water; and pipelines to distribute the water to where it is needed. A good deal of my energy and resources was devoted to developing the necessary farm infrastructure, especially in the early years.

Roads

The first requirement was a reliable road, and not having the suitable equipment, I hired that work done. The roadway was elevated and crowned with soil taken from adjacent barrow pits, and then covered with gravel (road base). Most of the gravel was mined locally, but some was asphalt grindings from nearby highway repair, and another portion was one-inch granite rock salvaged from an abandoned rail bed.

The graveled roads on the farm amount to about 1,600 linear feet, which required about 100 tons of gravel initially, and another 70 tons in maintenance over the years. In addition, there is three-quarters of a mile of common driveway shared with neighbors that was built at the same time, and which cumulatively has required several hundred tons of rock. Concrete used on the farm for the foundations of buildings, and which is mostly sand and gravel, amounts to another 110 tons.

Like all mining, gravel mining is not benign—it is a messy, destructive business harmful to both landscapes and waterways. In evaluating the ecological effects of a farming operation, gravel mined elsewhere constitutes a significant environmental cost. But in the case of roads and foundations, I have no good alternatives.

Buildings

I had worked in the building trades for years before becoming a farmer, and so I had the necessary skills—plumber, electrician, framer, roofer, cabinetmaker, plasterer—to build my own buildings without help. I am not a master of any of those trades, but I can do them adequately for my purposes. My wife, Dianne, was formerly an architectural draftsman; she has a very good eye, and the buildings are not shacks—they are handsome and functional. The greatest part of the cost of construction is labor, and by building with my own labor I was able to get useful buildings very inexpensively. The buildings, constructed over a period of years, are the house, studio, workshop, barn, greenhouse, packing and processing building, and guest house.

But it was not only labor where I was cutting expenses. Much of the construction used recycled materials. Although I had quite a bit of salvaged wood at the outset, it was not enough, and so I placed an ad in the newspaper offering to buy abandoned barns for salvage. A man called me up and offered a tumbling-down barn, which I bought for $100, and which provided me with thousands of feet of good lumber. Somehow, because of that, I began to develop a reputation as a person who dismantles old buildings, and people would seek me out, offering various old buildings that they wanted removed. For many years, I would take apart old buildings in the winter months when farm work was slow, and I would keep some lumber for myself, and sell the rest. The quality of lumber in 19th-century buildings of this region is remarkable; virgin timber was being cut, and sold cheaply, and used thoughtlessly. Some of the best salvage was too good for framing—I ended up building cabinets and furniture from it.

Another good source of salvage material came from roofing contractors. Many buildings in the region have roofs made of thick cedar shakes, and when just one shake goes bad, such a roof begins to leak. When the shakes are torn off of the leaky roof, they are mostly sold for two dollars a ton to a company that makes artificial fireplace logs. (For the roofer, the real motivation for this is to avoid paying high dumping fees at the county landfill.) I found that I could buy truckloads of old shakes inexpensively (a case of cold beer is a good starting offer). I would not put them back on a roof, but they are excellent for side-wall shingling, creating a handsome, rustic, durable (and flammable) surface.

After two decades of it, my building salvage work came to an end for a couple of reasons. One was that pulling nails with a crowbar all day long is very hard on the elbow joints, and I was developing chronic elbow pain. But the main reason was that what had been a low-budget business for down-and-out people like myself and my customers turned into a high-end business run by hipster capitalists. This was due in large part to LEED certification for environmentally exemplary construction, one requirement of which is the use of recycled materials. The price of used lumber rose quite rapidly, and competition for old buildings became intense, with extremely high prices being offered. This was another

The windmill no longer pumps water. The tower supports a bat house and parabolic dish for internet connection. Photovoltaic panels in the foreground power a direct current pump in the well.

case of the informal economy of poor people being commandeered by wealthier people.

The first building I built was my house, about 1,600 square feet with an encircling veranda—plenty of space for a family of four. But the idea of building a house at all might be questioned. Living on the farm is a peculiarity of the English-speaking countries. In the rest of the world, the farmer lives in the village and commutes to his farm. There are several advantages to this. His family is spared the social isolation of country life, and the considerable waste of time and energy in commuting to town for school, shopping, visiting, and other obligations. And the farmer has the luxury of knowing that when he comes home at the end of the day, his day's work is done. The farmer who lives on his farm always has unfinished chores before him, and goes back into the field after supper, or if it is dark, out to the shed to work on some equipment repair. In my case, with building lots and rentals in town being extremely expensive, this was not really an option for me, but under other circumstances I would surely have considered it.

Irrigation

As noted in the previous chapter, I have two small wells. The first, with a depth of 200 feet, proved insufficient during a severe drought in the early 1990s, and so the second well was bored, with a depth of 340 feet. I purchased an old water-pumping windmill from about 1920 and set it up over the original well after the aquifer had recovered, with the idea of having an extra water source. But the pump—an ancient copper cylinder and a piston with leather washers—never worked properly. The windmill remains, however. It is an interesting artifact in its own right, and the tall steel tower supports a bat house, occupied by several dozen bats, as well as a parabolic dish that mediates our connection to the internet. I can climb to the top of the tower for a good aerial view of the farm. In the meantime, the well is back in use with a direct current pump (Grundfos SQF) powered by five solar panels (1,400 watts). This supplies water for furrow irrigation in the orchards when it is needed to supplement drip irrigation.

Water is moved around the farm in buried PVC pipe, amounting to about 2,800 lineal feet of lines. There are seventy-two risers with valves, so that almost no cultivated part of the farm is far from a convenient source of water. Most of the final stage of irrigation is with drip tubing or drip tape, amounting to about 26,000 lineal feet (five miles). Formerly there was no local recycling option for retired drip lines, but now there is a facility that will take it. Managing these lines and keeping them repaired is a daily task during the irrigation season (March to October).

Electricity

We are connected to the grid by a long spur that reaches our farm and a neighboring one. A transformer on a pole supplies us with electricity at 240 volts and 60 hertz. Three-phase power is not available. Although the utility company is required to offer three-phase power if there is a legitimate need for it, they don't do it for free. The quote I got for a three-phase transformer was $34,000, to be followed by a monthly minimum charge of $200—an offer that I declined. The principal use of three-phase current would be for pumping water more efficiently and for operating my olive oil mill. I continue to use a less efficient single-phase pump for the water, and for the olive mill I installed a 25 kilowatt rotary phase converter to generate my own three-phase power.

We have a 3.5 kilowatt photovoltaic system on the roof of our house that is tied to the grid. This supplies about one-quarter of the electricity used on the farm. Another 1.4 kilowatts of photovoltaic panels off the grid power the direct current pump.

Chapter 8

Tools and Machines

The manufacture and use of tools is a defining trait of our species. Tools magnify our strength, increase our accuracy, and allow us to work difficult materials without injury. Farming, with its requirements of manipulating soil, water, and vegetation, is a tool-intensive enterprise.

The farmer of a few millennia ago made do with a flint knife, a pointed stick for planting, and the jawbone of an elk lashed to a wooden staff to serve as a sickle. Since the Industrial Revolution, the farmer's tool kit has increased greatly from this simple beginning. The sidebar on page 72 lists the tools and machines in use on my small farm. Collectively, they add up to about 12 tons (11,000 kg) of steel in an amazing variety of shapes and sizes. They represent centuries of accumulated human ingenuity and skill.

The Organization of Tools

My tools are organized according to the farming chores that might be addressed. There is a toolbox for plumbing repair, a toolbox for electrical work, a toolbox for engine work, a rolling tool cabinet with tools for the olive mill, a pushcart dedicated to irrigation tools, and so forth. Most of the woodworking tools are stored in the workshop; the pressure washers and tools with gasoline engines are in the barn; the tractors and van sit under

Tools and Machines in Use on the Farm

1 half-ton cargo van
1 48-volt electric flatbed truck
2 diesel tractors
1 gasoline tractor
2 bicycles
5 push carts
1 wheelbarrow
1 hand truck
1 chisel plow
1 disk harrow
1 spring-shank harrow
1 ripper
1 smooth roller
1 ring roller (cultipacker)
1 flail mower
1 tractor-mounted rototiller
1 tractor-mounted brush chipper
1 tractor-mounted rake
1 box scraper
1 bed shaper
2 cultivating sleds
1 flatbed trailer
1 tank trailer
1 8-foot float
1 20-ton wood splitter
1 half-ton per hour olive oil mill
2 grease guns
5 oil cans
1 engine hoist
1 lawnmower
7 spades
6 shovels
9 hoes
2 pitch forks
3 rakes
2 pick-axes
3 cultivators
2 mauls
2 scythes
6 wire cutters
3 metal shears
1 conduit bender
3 vises
1 anvil
6 crowbars
4 soldering irons
16 bar clamps
14 C-clamps
8 spring clamps
1 drill press
4 electric hand drills
4 human-powered drills
107 drill bits
5 socket wrenches
80 sockets
38 box wrenches
30 allen wrenches
41 screwdrivers
1 gear puller
1 table saw

1 band saw
2 chop saws
1 worm-drive saw
6 sickles
1 post hole digger
1 post puller
1 sledgehammer
1 gasoline-powered cultivator
1 cold water pressure washer
1 hot water pressure washer
1 electric compressor
1 gasoline-powered compressor
2 air chucks
1 pneumatic blower
1 pneumatic lopper
1 pneumatic olive rake
1 pneumatic metal shear
1 gasoline string trimmer
1 gasoline hedge shear
2 hand shears
5 secateurs
5 scissors
9 loppers
4 harvest knives
17 ladders
5 chainsaws
49 hand saws
22 wood planes
6 spirit levels
17 steel files
4 spoke shaves
11 nail sets
1 nail puller
6 drift pins
3 staple guns
16 hammers
4 mallets
42 chisels and gouges
6 cold chisels
9 pipe wrenches
2 pipe cutters
1 bolt cutter
2 jig saws
3 electric grinders
1 jointer
1 planer
1 shaper
18 shaper knives
2 belt sanders
11 pliers
9 tape measures
7 calipers
8 squares
2 hydraulic jacks
1 mechanical jack
1 tile cutter
1 tile saw
6 concrete finishing tools
6 plaster floats
2 electric scales
1 spring scale
2 soap blenders
1 heat gun
1 drip tubing coiler
1 drip tape reel
500 other miscellaneous small tools

shade trees; the implements are parked in the open; and the electric truck is kept in a shed. Each tool has a place, and is either in its place or in use.

A perusal of the table suggestsquite a bit of redundancy. For example, there are 17 ladders. Six are orchard ladders, seven are stepladders, the rest are lean-to ladders and an extension ladder. The stepladders are distributed among the buildings so that there is always one nearby, and you don't have to fetch a ladder from 500 feet away when you need one, and then return it when you're finished with it. Similarly, each toolbox has three sizes of Phillips screwdrivers. If I grab my plumbing repair toolbox, I know for certain that I will have the screwdrivers I need, and I won't waste time searching other toolboxes for them. Redundancy is a critical element of efficiency. With half a dozen tape measures in the workshop I can always locate one in less than a second. If I had only one tape measure and it wound up out of sight under a board, I might waste valuable minutes searching for it.

Three tractors might seem excessive on a one-man farm, but they are of different design and purpose. The old 1947 gasoline tractor is rigged for cultivating. The four-wheel-drive diesel is my soil-working tractor. The other diesel (two-wheel drive) I use only for mowing. All the tractors are used for pulling flatbed or tank trailers from time to time.

A large number of tools in a category does not necessarily imply redundancy, however. Of the sixteen hammers, no two are the same. They range in weight from a delicate 4-ounce tack hammer to a 40-ounce hammer for heavy pounding. There are ball peen and cross peen hammers for use on metals, and claw hammers for driving nails. Depending on the size of the nails, the density of the wood, and the strength of your arm, a 16-, 18-, 20-, 22-, or 24-ounce hammer might be the most appropriate for driving nails. To one unfamiliar with tools, it might seem that a hammer is a hammer, and a ladder is a ladder, and a screwdriver is a screwdriver, and that's all there is to it, but he would be mistaken.

Seemingly similar tools have subtle but important differences, and usually only one tool is the correct one for a job.

The Care of Tools

Taking care of tools is one of the most profitable uses a farmer can make of his time. Keeping them in their proper places and out of the weather is part of this. The most important practice is the lubrication of moving parts. In the absence of lubrication we have noise, abrasion, friction, heat, and a short life. Put in a good layer of lubricant and we have quiet, smoothness, coolth, and easy running. The oil can and grease gun can hardly be overused.

Wooden handles get a good soaking with linseed oil during the dry season, and perhaps a coat of wax where the hand contacts the handle. Cutting tools benefit from frequent sharpening; when I'm pruning in the orchard I sharpen the loppers three or four times a day. Just one or two passes of a stone on the cutting edge suffices. Tools that move through the soil are self-sharpening, as the tiny mineral particles of the soil steadily scrape away steel. An old, well-used spade becomes thin and sharp as a razor, which is why an old spade is a better tool than a new one.

The Cost of Tools

Some of my tools I inherited. The cost of tools and machines that I purchased, not adjusted for inflation, is about $135,000. This is greater than the cost of the buildings on the farm. The most expensive items were the olive oil mill ($72,000 in 2005), the diesel tractors ($14,000 each, one purchased new, one used), and the cargo van ($6,500 used).

As with other manufactured goods, the price of tools, adjusted for inflation, is radically lower than it was a few decades ago. I purchased a set of box-end wrenches, made in the US, for $20 in 1972. Adjusting for inflation, that would be $112 in the currency of 2015. I could now buy the same set of wrenches, of equal quality, made in China, for about $12, or nearly one-tenth of the inflation-adjusted price I paid in 1972. How is this possible?

Partly the lower price represents outsourcing of production to places with a lower cost of labor. And partly the lower price reflects increased efficiency and increased automation in production, including mining, smelting, manufacture, shipping and distribution. But there is an ominous side to low prices as well; what economists call "negative externalities" are costs of the item not accounted for in its price. In the case of the wrenches mentioned above, these costs include gross air pollution and water pollution in China with their negative consequences for public health and the environment; pollution of the oceans and damage to fisheries cause by trans-oceanic shipping; unemployment and its social consequences in the US; oil wars, bad foreign policies, and climate change resulting from the US insistence on the cheap price for fossil fuels that underpins the global economy; damage to North American aquifers as a result of fracking; and numerous similar consequences. In this context, purchasing a tool is not a morally neutral act, and the utility of the tool must be weighed against the social and environmental costs of excessive manufacturing and commerce. Redundancy of tool ownership can improve efficiency of work on the farm, but it also has a negative side.

Intensity of Use and Lifespan of Tools

The manufacture and distribution of a tool has an ecological cost. Fossil fuels and materials are used up, and there are negative ecological aspects to mining, smelting, manufacture, and shipping. These ecological costs should be balanced by the utility of the tool. The manufacture of a tool is most fully justified if the tool is heavily used until it is fully used up (recognizing that some tools, such as hammers and wrenches, are potentially immortal). However, it is intrinsic to farming that most of the tools are idle most of the time.

On my farm, the olive mill is idle for 43 weeks of the year, and the bed shaper gets only about ten hours of use in a year. Many of the other implements are used for only a few days, or even just a few hours, per year. Some of the hand tools may be called on only once in five years. If you study the market for used farm equipment or medium-sized trucks, it is shocking how little use most machines have received in the course of fifty or sixty years.

The expected working life of a medium-sized tractor is about 6,000 hours of use. One of my tractors I use almost exclusively for mowing; it gets about 50 hours of use per year, which means that it should last 120 years. But the aging of machinery is not only a question of use; just sitting in a shed, the machine also ages. Oxygen, and oxygen's big brother, ozone, chip away at seals and gaskets and paint, especially in the presence of sunlight. Steel rusts, aluminum oxidizes, and rubber hardens and becomes brittle. Moreover, technology continues to advance, so that the old machine with its antiquated technologies is left behind. The tolerances to which engines were built fifty years ago are very sloppy by today's standards; complicated and inefficient carburetion has been replaced by fuel injection; drum brakes are replaced by disk brakes; crude ignitions are replaced by precise computerized systems. The old machine may still function, but it is loud and uncomfortable and undependable, and it is a gross polluter.

From an agroecological perspective, there is a tension between the increased efficiency that derives from owning a wide range of tools and the ecological cost of those tools, and in this dialectic, small farms are at a disadvantage. On a big farm, a tractor and disk may work for 1,500 hours a year, and in a few years they are fully used up and their ecological costs of manufacture have been fully amortized by the useful work they have accomplished. But on a small farm, where the tractor and disk are used for only 100 hours per year, the amortization is never completed. And, as Karl Marx pointed out (*Capital*, Part IV, Chapter 15), assuming fixed lifetime output, the faster a machine is used up, the greater the annualized return on investment.

An obvious solution for small farmers is to share equipment, an appealing idea that almost never works in practice. Partly this is because of differences in personality that lead to disagreement and unhappiness. The orderly partner takes good care of the equipment, and the sloppy partner abuses it and returns it with the fuel tank empty. Or the partners have equal investment, but one partner uses the equipment twice as much as the other. But even if the partners are compatible, it is the natural cadence of farming that they both need the equipment at exactly the same time. The winter rains have ended, a north wind comes up and dries out the soil, and now is a critical time for spring plowing when the soil moisture content is perfect—in another few days it will be too dry. Both partners need that equipment today, even if it has sat idle for the last four months and will be idle again by next week. On a regional basis, it is this time sensitivity of farming practices that underlies the redundancy and underutilization of equipment.

For tools that are seldom used, rental from a rental yard is a sensible approach. I rent a trenching machine if I need to dig a long trench for a pipe or cable, and I rent an equipment trailer if I need to move a tractor. For bulk materials like compost or gravel, I hire delivery rather than owning or renting a suitable large truck. But for the equipment routinely used on the farm, the need for efficiency dictates that I own a lot of tools, even though they are underutilized.

Chapter 9

Energy

In one sense, the farmer's business is energy. The aim of his farming strategy is to divert as much of the sun's energy as possible into useful chemical energy stored in his crops. He achieves this by his knowledge of how to manipulate the environment, by his own labor and the labor of his draft animals, and by the use of fossil fuels. Energy from the sun is essentially perpetual and reliable, as are the mechanical energies of men and animals, whereas fossil fuels are a finite and rapidly diminishing resource. And so the efficiency with which fossil fuel energy is used to amplify the capture of solar energy is a major ecological concern. Note that the role of fossil fuel was unimportant 100 years ago, and may be unimportant again 100 years hence, but at present it is a critical issue for farming. The chemical energy content of the harvested crop divided by the input of fossil fuels used to produce the crop defines an energy efficiency ratio. (A slightly different definition of the efficiency ratio might include the input of renewable energy, such as the labor of humans and draft animals; or wind, photovoltaic, or hydroelectric energies.)

Seldom is the farmer concerned with the energy efficiency ratio as an ecological measure; usually his concern is primarily economic. Other things being equal, a more efficient farming system may also be more profitable, but this is not always the case. If fossil fuels are inexpensive in relation to the value of the crop, a less efficient system with a higher overall yield may be more profitable, and in that situation most farmers will choose greater profit over greater efficiency.

The measurement of an energy efficiency ratio would seem to be a straightforward calculation; you measure the energy used to produce the crop and the energy content of the crop, and there is your ratio. But these measurements have fuzzy edges. For example, suppose you have twenty tons of compost delivered as a soil amendment. The chemical bonds of the organic molecules making up the compost are the result of solar energy. But fossil fuels were used at the compost manufacturing facility to power the grinders and hammer mills and conveyors and compost-turners. And the truck that delivered the compost used fossil fuels to transport it. Not only that, but fossil fuels were used to manufacture the truck and the compost-making machinery, and some increment of that energy must be attributed to the compost that was delivered. And even further back, the energy used to build and operate the steel mill that made the steel of which the truck is built also must have some tiny amount attributed to this load of compost. If the input material of the compost is urban green waste and wood scraps, it is likely that irrigation, fertilization, gardening machines, and forestry machines made contributions to the compost, also using fossil fuels. So what is the fossil fuel input of twenty tons of compost? It's not easy to say.

The measurement of human and animal labor is also imprecise. We can measure the work done by a draft animal in plowing, but should we not also consider the energy used by the animal when it is at rest, or walking from its labors back to its shed, which after all, is part of the labor of plowing? It might make more sense to measure the energy content of the feed going to the animal over its lifetime, including those periods when it was too young or too old to work. And by analogy, should we not include the energy used by humans when they are asleep or engaged in household tasks unrelated to farming? While there is not a right or wrong answer to questions of this type, it is important that the limits of the analysis are carefully specified.

The energy content of the harvested crop is more readily and clearly measured than are the inputs of fossil fuels. But from the perspective of the human users of the crops, not all energies are equal. One megajoule of sucrose has the same energy as one megajoule of mixed salad greens and one megajoule of beefsteak, but these are very different foods with highly different values. It may be more energy-efficient to make sucrose than to make salad, but that does not mean that sucrose is a superior food.

There is not necessarily a right or a wrong way to calculate energy efficiency, but it is critical to specify how the measurements are made, and what is included and what is excluded. Because various authors have used different criteria and different measures, the published energy efficiency ratios for various crops are often not directly comparable. There is a scholastic tradition of including human labor in calculating energy inputs, but from an ecological perspective the critical issue is fossil fuels, which constitute the only nonrenewable energy input.

Solar Energy

Of the energy reaching us from the sun, we are most interested in wavelengths that we perceive as light, which drive photosynthesis, and the longer wavelengths that we perceive as heat, which help to create the thermal conditions conducive to plant growth. The farther one is from the equator, the more that the annual flux of solar radiation varies throughout the year. We see this as short day length, a low angle of the sun, and less heat in winter, with longer days, a steeper sun angle, and more heat in summer. In addition to these factors, annual variations in cloud cover may also affect solar radiation. For example, some regions around the Great Lakes are cloudy throughout the summer, and the diminished light intensity may suppress crop productivity.

While day length and net solar energy are greatest at the summer solstice, at my location there is a lag in air temperature, which peaks in mid-July, and soil temperature, which peaks in late July or early August. Those delays reflect the situation of the earth slowly heating up and slowly drying out.

Photosynthesis is not a highly efficient user of solar radiation; the net increment is typically less than one percent. Efficiency is greatest when the leaf area index is between about three and seven, i.e., the total area of the leaves above a one square meter plot of ground amounts to between three and seven square meters. The tree crops—olives, figs, citrus, and apricots—are within this range throughout the summer; persimmons have a higher leaf area index. To maximize crop growth, one would want to achieve a suitable leaf index by the time that solar radiation is at its peak, around the summer solstice.

If I climb up the tower of my old windmill, from where I can get a good view of my farm, I will see, on the summer solstice, that much of the open ground has no plant growth on it. These areas were planted to winter-grown crops that have already been harvested. By late August, when there is still abundant solar radiation and heat, almost all of the open ground is bare. What this signifies is that my farming system does not make good use of incoming solar radiation. The reason is that this is a water-limited farming system, rather than a light-limited or heat-limited system. And even in the orchards, which have an optimal cover of leaves, photosynthesis can be inhibited by an insufficiency of water, which undoubtedly happens on my farm. If we were to measure efficiency as the percent of incoming solar radiation that is used to produce plant growth, then the farm would register well below its potential.

Energy Inputs

Traditionally, energy is measured using a variety of units: Btu, kilowatt-hours, calories, and so forth. To simplify comparisons, and to remain consistent with published literature, I present all of the energy measurements in units of megajoules (MJ). (I should mention that I have not rounded numbers to significant digits, but simply let the calculations come out as they will. The reader should recognize that most of the numbers are estimates that are not as precise as their presentation might suggest.)

The annual input of human labor on the farm is presented in table 9.1. Immediately some of the issues mentioned earlier become apparent. For example, 810 hours spent on sales (mostly farmers markets) take place off the farm and in one sense are not an input that should be measured in calculating efficiency; yet selling the products of the farm is clearly part of the human labor of farming, as is post-harvest transformation—turning olives into olive oil and apricots into apricot jam, which is very time-consuming. At any rate, I choose to measure all labor, including that which is not spent producing crops but processing them, selling them, and taking care of the necessary paperwork. Not included are the activities of daily life unrelated to these, such as cooking, housekeeping, reading, and visiting, even though they are a part of farm life.

Olive harvest and tree pruning are demanding, aerobic work of the highest order, and I assign to them an energy use of 1.1 megajoules (MJ) per hour. The other activities listed in the table are less physically demanding, and are valued at an energy expenditure of 0.6 MJ per hour. This gives a value for the total energy of human labor of 4,049 MJ per year. These measures of energy use reflect the metabolic rate of the worker while working, rather than the work accomplished.

Fossil fuels used in farming include diesel fuel to power tractors and as a heat source for a hot-water pressure washer; gasoline for chainsaws, mowers, and the old cultivating tractor; and propane used for heating water in the packing shed and for cooking fruit to make jam. The amounts used per annum are: diesel fuel, 75 gallons (10,170 MJ); gasoline, 11 gallons (1,346 MJ); and propane, 70 gallons (6,230 MJ). The total is 17,746 MJ. Note what has not been included here: all of the driving of vehicles involved in purchasing and in sales, which would be approximately 400 gallons of gasoline

Table 9.1. Farm Labor Through the Year (Two Adults)

TYPE OF WORK	HOURS
Olive harvest	710
Pruning	340
Weeding	260
Irrigation	200
Crop production/harvest	1,740
Post-harvest value added	1,600
Sales	810
Purchasing/logistics	65
Infrastructure maintenance/construction	220
Equipment maintenance	50
Paperwork	40
Teaching/mentoring and conferences	38
TOTAL	6,073 hours

(48,960 MJ) per annum. So the off-farm driving in logistic support of the farm consumes three times as much fossil fuel as the on-site operation. Moreover, there are other fossil fuel expenditures that can be attributed to the farm; for example, when the UPS truck comes down the road to deliver a shipment of seeds, not only is the delivery truck using fossil fuel, but other vehicles that transported the seeds from their point of origin in another state must also have some portion of their fuel use assigned to the farm. Similarly, a large truck delivering compost from a facility 25 miles distant is burning fossil fuel that should be considered part of the energy input of the farm, both the delivery with the truck loaded as well as its return empty to its point of origin.

The use of fossil fuels could be eliminated from the farm by substituting olive oil, which could power the tractors and stove with little modification, and which could be traded for an energy-equivalent amount of gasoline for the chainsaws. This could be achieved using less than 15 percent of the olive oil production, and as the trees mature and production rises, that figure would fall below 10 percent.

Electricity, both from the grid and from the two photovoltaic systems, is used to power the cooling machinery of the walk-in cooler, the olive oil mill, the irrigation pumps, and the small electric truck, as well as lights and electric tools. Because both farm use and household use of electricity are routed through the same meter, and because the off-grid photovoltaic water pumping system is not metered at all, the values for total electricity use are approximate. What I find is a total annual use of 59,040 MJ, of which about 27 percent is from photovoltaic systems and the remainder is from the grid. If I apportion 20 percent of the total to non-farm household use, then the net farm use is 47,230 MJ.

The local utility from which I buy electricity claims their sources as follows: 21 percent nuclear power, 11 percent hydroelectric, 19 percent wind and solar, and the balance from fossil fuels, primarily natural gas.

In conventional agriculture, a major energy input is synthetic fertilizers and pesticides, the manufacture of which requires high use of fossil fuels. For example, most nitrogen fertilizers are manufactured from atmospheric nitrogen and natural gas (methane), with the natural gas serving as both an energy source and a material input. One thousand pounds of nitrogen in synthetic fertilizer uses 1,800 pounds of fossil fuel in its production. Because I don't use synthetic fertilizers or pesticides, this energy use does not apply.

Some energy analyses of agricultural systems neglect the embodied energy of the farm equipment. "Embodied energy" refers to the energy required to manufacture and distribute the equipment, as well as the energy required to dispose of it when its useful life is over. Embodied energy can constitute a significant portion of the farm's energy budget, and so I will make an attempt to calculate it. Approximately 11,000 kg of steel make up the equipment (tractors, implements, hand tools, cargo van) used on the farm. The embodied energy of raw steel is approximately 20.1 MJ per kg. But to turn 2,000 kg of steel into a 2,000 kg tractor requires considerably more energy for fabrication, assembly, and distribution. I do not have good data that measures this, but it seems reasonable to assume that this doubles the embodied energy, raising it to about 40 MJ per kg. So the 11,000 kg of equipment represents an embodied energy of 440,000 MJ. We must divide this by the annual increment of the lifespan of the equipment in order to calculate the annual value. If I assume, conservatively, that steel tools and machines have a lifespan of 40 years (some of them are already a good deal older than that), then the annual charge to embodied energy is 11,000 MJ.

The farm has 34 square meters of photovoltaic panels, with an embodied energy of 4,070 MJ/m^2, for a total of 138,380 MJ. Dividing this by a lifespan of 25 years gives an estimate of 5,535 MJ per annum. Slightly more than a third of the energy output of the panels must be used to underwrite the cost of their embodied energy. This figure is an agreement with the calculations made by Prieto and Hall (2012), although others (e.g., Ugo Bardi) calculate the embodied energy of the panels at less than 10 percent of their lifetime output.

Calculation of the embodied energy of the farm buildings is obscured by the consideration that they are largely built of recycled materials that have already served their useful life in some other structure. But I can at least estimate embodied energy of the concrete foundations (this includes the packing shed, workshop, and barn, but not the dwellings and studio).

Approximately 48,000 kg of concrete were used, with an embodied energy of 1.1 MJ per kg, for a total of 52,800 MJ. Dividing this by an expected lifespan of 50 years gives an annual energy input of 1,056 MJ.

Table 9.2. Annual Energy Inputs to the Farm

ENERGY SOURCES IN MJ		PERCENTAGE SUMMARY	
Human labor	4,049	Human labor	5%
Diesel, gasoline, propane	17,746	Fossil fuel	40%
On-farm photovoltaic	12,752	Nuclear	8%
Grid: nuclear	7,240	Hydroelectric	4%
Grid: hydroelectric	3,793	Wind and solar	22%
Grid: wind and solar	6,551	Embodied energy	20%
Grid: fossil fuel	16,894		
Embodied energy	17,591		
TOTAL	**86,616 MJ**		

A summary of the annual energy inputs to the farm is presented in table 9.2. It is notable that despite long hours of hard work, the energy supplied by human labor represents only about 5 percent of the total. Looked at another way, if all of the energy requirements were met by human labor, it would require 40 workers putting in 60-hour weeks throughout the year to meet the demands of the farming operation.

Energy Production

A summary of energy production as measured by the energy content of the harvested crops is presented in table 9.3. By far the greatest energy production on the farm is to be found in olive oil. Olive oil is an energy-dense material, and olives comprise half of the cultivated ground. The harvest varies from year to year, but averages about 3,100 liters of oil. Of the olive trees, 200 are newly planted and not yet bearing. Another 180 trees were burned to the ground in a

Table 9.3. Farm Energy Production

CROP	AMOUNT HARVESTED	ENERGY IN MJ
Olive oil	3,100 liters	114,731
Melons	3,200 kg	3,744
Tree fruits & berries	1,200 kg	2,400
Flowers	4,000 kg	8,000
Garden vegetables	1,200 kg	2,400
Firewood	1 cord	22,300
	Total, harvested crops	**153,575 MJ**
	Annual increase in standing biomass	**133,910 MJ**
	Total	**287,485 MJ**

fire; they have sprouted from the roots, but are not yet bearing fruit again. A further 160 trees are five years old and are bearing only lightly; their productivity should increase over the next few years. As the grove comes into full maturity, the average yield should rise to about 5,000 liters per annum, with no increase in energy inputs.

Melons have a high water content and a correspondingly low content of energy so that, despite the considerable weight of melons harvested, their contribution to energy production is low. For fruits, vegetables, and flowers other than melons I have assigned an energy value of 2 MJ per kg fresh weight; the literature gives variable estimates, but most are in this neighborhood.

My house is heated by firewood burned in an iron stove. In this mild climate the total use is about one cord (3.62 cubic meters) per year, mostly cut from limbs of the fruit trees (olive, apricot) during routine pruning. If an old eucalyptus tree falls, that also is cut for firewood. Olive and eucalyptus are both woods with high energy density.

While the usual convention of energy analysis in agricultural systems is to measure only the energy content of the harvested crop, it strikes me as worthwhile to also note the energy embedded in the increasing biomass of the farm. What had been open ground at the outset is now a cultivated forest of trees, some of them quite large, and bearing large root systems. This increase in the chemical energy of wood and leaves is due in large part to the energy inputs (compost, pruning, orchard floor management, and—most of all—irrigation) inherent in farming. Moreover, carbon sequestration in the soil has been greatly increased by the practice of piling the pruned branches into windrows and converting them to chips with a chipping machine; the chips may be left as a mulch, or shanked into the soil if soil compaction requires cultivation. A rough estimate of the energy effect of this is as follows: for 1,913 fruit trees, assume a modest 5 kg per year increase in dry weight of wood with an energy density of 14 MJ per kg, giving an annual increase of 133,910 MJ.

Analysis

In the agroecological literature, increased residual biomass as an output is usually not included. Leaving that factor out of the calculation, for my farm I find a ratio of 2.94 units of energy produced (i.e., harvested crops) per 1.0 unit of energy input. Note that this input is for fossil fuels and embedded energy (which is mostly fossil fuels) only, and does not include renewable energy inputs (photovoltaic, wind, hydroelectric, and human labor). If renewable energy inputs are included as well, the efficiency ratio drops to 2.1. To put this in context, calculations for other crops conventionally farmed in North America include dryland wheat (11.13), corn (3.89), soybeans (3.19), rice (2.2), potatoes (1.3), oranges (1.0), and processing tomatoes (0.26). Studies of swidden agriculture (New Guinea) and semi-industrialized agriculture (India, Polynesia) give values in the range of 10 to 14; clearly, industrialization of agriculture has decreased its energy efficiency while increasing total yield. Much of the increase in yield must be interpreted as turning fossil fuels into food.

The numbers cited in the previous paragraph are not readily comparable because of the ways in which they are calculated. For example, the very high efficiency for dryland wheat calculated by Spaeth (1987) does not include harvest of the crop, drying, or post-harvest transport and processing; including those inputs would decrease the calculated efficiency. The embodied energy of machinery is sometimes estimated using the lifespan provided by depreciation schedules of the Internal Revenue Service, in which machinery is considered to be entirely used up in seven years. If one accepts this, then the rate of embodied energy use becomes sensitive to scale, with the per acre input less on bigger farms.

My calculated ratio (using only fossil fuels as inputs) of 2.94:1 includes the energy input of post-harvest processing: the energy-intensive operation of a walk-in cooler and the operation of the olive oil mill. Such post-harvest energy demands are not always included in the calculation of efficiency ratios.

When increased biomass as an energy output is added to the calculation, the ratio of energy produced to energy input rises to 5.51:1. Much of the applied energy, especially that of irrigation, is stored in the form of growing orchard trees and increased soil organic matter. If this analysis were applied to annual crops such as corn and wheat, the culture of which may lead to a decrease in net biomass, especially if the straw is harvested, their energy efficiency ratios would be less than what has been published.

I have classified electricity produced on the farm using photo-voltaic panels as an energy input. Clearly the embodied energy of the system is a cost, but one could argue that the energy produced is a product of the farm rather than an expense. A middle path might be reasonable: to acknowledge the embodied energy as an input, and to simply ignore the energy produced rather than treating it as either an input or an output. Its value will show up elsewhere in the calculations, since the amount of energy purchased from the grid will be correspondingly diminished.

For modern farming in North America, probably the major determinant of the energy efficiency ratio is the type of crop grown. Grains are annual plants that put nearly all of their energy into their seeds, so that they have a high ratio of harvested energy and a low residuum of vegetative mass. Tree fruit crops, on the other hand, are perennials, and only a minor portion of their annual captured energy is devoted to making fruits, with the balance devoted to increasing the vegetative mass of the plant. Cantaloupes and watermelons, which we treat as annuals, are short-lived perennials in warmer climates. They are adapted to desert situations in which sequestration of water (rather than energy) in the fruit is the strategy for attracting animals who would disperse seeds. In an agricultural setting, the annual fruit crops (melons, tomatoes), which are more than 90 percent water, require considerable energy for harvest simply because of the weight of water that must be moved.

Most of the crops grown in my region have an energy efficiency ratio of less than one. The dominant crops are tomatoes (0.26) and walnuts (0.81). For walnuts, the critical input is irrigation; 25 inches or more of irrigation pumped from groundwater and distributed through sprinklers at high water pressure is very energy-demanding. The reason that the crop is financially viable with an energy ratio of less than one is that the input energy costs less than 10 cents per MJ, while the harvested energy is valued at well over one dollar per MJ.

The European Union recently published a lengthy document describing practices that could be applied to increase the energy efficiency of agriculture. Some of these are relevant to my operation, such as using equipment of appropriate horsepower (many tractors in this region are overpowered in relation to their tasks); driving correctly (optimal engine speed, no idling); and using controlled traffic (e.g., permanent raised beds). However, the most likely route to increasing efficiency is to increase yield with the same inputs. As the olive trees come into maturity, I expect yield to rise by about 60 percent. But even more valuable would be an increase in the thoroughness of harvest—a topic covered in the next chapter.

Carbon Footprint

With the exception of a portion of the electricity, all of the energy used or created on the farm involves reduction or oxidation of carbon compounds. Because of this, the carbon economy of the farm is very similar to the energy economy. There are a few differences, however. For example, carbon stored in the soil as a carbonate (e.g., magnesium carbonate) counts as sequestered carbon, but it has no energy value, being already oxidized.

The amount of fossil fuel used by an individual, or a household, is in some sense a measure of global good citizenship. For a resident of the US, the environmental and social costs of using fossil fuels include bad foreign policies, oil wars, a costly military patrolling the seas, tax giveaways to oil producers, air pollution and water pollution (with associated threats to public health and ecosystem health), and acceleration of climate change. Most of these costs are not reflected in the price of fossil fuels, thereby creating a false impression of the economic consequences of using fossil fuels. Looking beyond the price of fuel to its true cost reveals an inescapable ethical imperative to reduce one's carbon footprint.

Simply living on the farm forces an increase in the use of fossil fuels for the activities of daily living. When our children were younger, we sometimes made four or five car trips to town in a day, meeting the requirements of school, music lessons, soccer practice, dental appointments, grocery shopping, and the like. And even now that our children are adults and no longer live with us, we still, between the two of us, make about 400 trips a year to town in the course of running a household and a farm, and in keeping up with friends. If we lived in town, these trips could be made on foot or by bicycle. So there is a compelling argument that an extra 6,000 to 10,000 miles a year of driving a car is part of the energy cost of farming. Our farm is seven miles from town. We know farmers who live more than fifty miles from town, and who log 40,000 miles a year or more on the road just taking care of daily life. Even though this travel is not directly a part of farming, it is a consequence of living on the farm, and should be included in a full energy analysis. Note that this variable is sensitive to scale; the additional energy input per acre consequent to living on the farm rather than in town is much less on a 1,000-acre farm than on a 10-acre farm.

For the farmer, the estimate of a carbon footprint includes the uses of fossil fuels and the carbon value of embedded energy of tools and buildings. But in addition to these, and often of much greater magnitude, are the changes in the carbon content of the soil and the aboveground biomass of the farm. A farmer growing annual crops with frequent tillage and high inputs of anhydrous ammonia may be causing significant loss of historic soil carbon. Conversely, a farmer raising perennial crops with minimal tillage may be sequestering considerably more carbon than is expended in farm operations.

The Significance of Energy Efficiency

Consider the case of a foolish youth who, at age 21, inherits a fortune that he spends so recklessly that, by the age of 30, the fortune is dissipated and he finds himself destitute. This is more or less the situation of the human species. We have inherited great wealth in several forms: historic solar energy, either recent sunlight stored as biomass, or ancient sunlight stored as fossil fuels; the great diversity of plants and animals, organized into robust ecosystems; ancient aquifers; and the earth's soil, which is the basis for all terrestrial life. We might mention a fifth form of wealth—antibiotics, that magic against many diseases—which we are rendering ineffective through misuse. Of these forms of wealth that we are spending so recklessly, fossil fuels are primary, because it is their energy that drives the destruction of the other assets.

What we have purchased with the expenditure of this inheritance is an increase in the human population of the planet far above what the carrying capacity would be without the use of fossil fuels. This level of population cannot be sustained, and so must decline. The decline could be gradual and relatively painless, as we see in Japan, where the death rate slightly exceeds the birth rate. Or the decline could be sudden and catastrophic, with unimaginable grief and misery.

In this context, the value of increased energy efficiency is that it delays the inevitable reckoning; that is, it buys us time. We could use this time wisely, to decrease our populations in the Japanese style, and to conserve our soil, water, and biological resources. A slower pace of climate change could allow biological and ecological adaptations. At the same time we could develop and enhance our uses of geothermal, nuclear, and solar energies, and change our habits to be less materialistic. A darker option is to use the advantages of increased energy efficiency to increase the human population even further, ensuring increasing planetary poverty and an even more grievous demise. History does not inspire optimism; nonetheless, the ethical imperative remains to farm as efficiently as one is able.

The tragic side of this situation is not so much the fate of the humans; we are a flawed species unable to make good use of the wisdom available to us, and we have earned our unhappy destiny by our foolishness. It is the other species on the planet, whose destinies are tied to ours, that suffer a tragic outcome.

Chapter 10

Productivity

Harvest on the farm continues throughout the year, and there is no week when we are not harvesting something. In January we are still harvesting olives, as well as citrus, and by the end of the month the earliest flower crops are coming in. February through June is the most intense season for harvesting flowers, along with citrus; blackberries are ready in May and June, and apricots ripen in June and July. July through September we continue with flowers, figs, watermelons, and cantaloupes. By October flowers are winding down as harvest of persimmons and olives ramps up.

Incompleteness of the Harvest

A modern combine harvesting a well-grown field of wheat will secure 99 percent of the crop, with only a few grains spilled on the ground. Harvesters of oranges and avocados also aim for 100 percent. But for soft tree fruits such as apricots, persimmons, plums, figs, and olives, complete harvest is not a reasonable goal. It is possible to harvest 80 percent of the fruit fairly quickly, but then the pace slows down. Some fruit is borne in hard-to-reach locations in the interior of the tree; a few fruits will be out at the tip of a tall branch where it's hardly worth moving a ladder to get to them. Some of the fruit will be damaged by wildlife or hail, or will still be immature at the time the tree is harvested, and these fruits will be left on the tree or knocked to the ground.

On my farm, the thoroughness of harvest is, on the average, very poor. For the olives, I harvest about 50 percent of the fruit, sometimes less. The production of the orchard is about 25 to 50 tons, whereas the harvest is 12 to 30 tons, with an average of about 19 tons. Early in the season, the 80 percent rate of harvest is often achieved, but as the season progresses, this diminishes. As the fruit ripens, an abscission layer forms at the base of each olive, which allows it to be easily removed from the tree. Often a storm with high winds will knock half of the fruit off the trees. With half the fruit on the ground, the net harvest rate per tree now declines to about 40 percent with the same input of labor that previously yielded 80 percent. A hard freeze can damage the fruit, ending the harvest prematurely. And cumulative fatigue of the worker can lead to an end of the harvest before all of the fruit is picked.

Olive harvest starts in mid-October, and runs until early January. Risks to the crop (high winds, freeze) occur toward the end of that season, and so an obvious way to improve the efficiency is to accelerate the harvest so that it is completed by mid-November. Indeed, this is what many growers do. There are two ways to do this. One is to mechanize harvest with a mechanical harvester (tree shaker or limb shaker). The purchase price ($80,000 to $200,000) is more or less balanced by the cost of forfeiting part of the crop if one lacks mechanization. If I were twenty years younger, I might opt for mechanization, but this late in my career it doesn't make economic sense.

Another route to mechanization would be to hire a harvesting contractor who has appropriate machinery. However, the capacity of my mill is sized to hand-harvest. A mechanical harvester could easily harvest 15 tons of olives per day, but my mill can only handle about two to three tons per day. So I would have to find a contractor willing to bring his equipment in and then work for only a part of each day for two weeks.

The other way to accelerate the olive harvest is to hire help. I tried this one year, registering with the tax authorities and taking out a policy of worker's compensation insurance. The paperwork was intolerable, and the insurance company was sending me bills every week for all sorts of abstruse charges. They required large cash deposits, and then charged me management fees for holding them. Filing tax papers led to another nest of problems, as some of the Mexican workers' names did not match their Social Security numbers, and I got threatening comments from the tax authorities. At the end of that season, I returned to my status as a solo worker with no employees.

An alternative is to use a labor contractor to shoulder the bureaucratic burden. Locally, the contractor's charge is 42 percent of wages. So if I pay a worker $11 per hour, the cost to me is $15.64. At this rate, the wages almost cancel out the advantage of increased harvest. Other growers use hired workers through a labor contractor, but they price their final product much higher than I do, which enables them to absorb that cost. The issue of pricing will come up in the next chapter.

Many farmers exploit unpaid or poorly paid interns to help out when labor demands are high. I find that to be unethical. When I have interns on my farm, I offer them the use of land, water, equipment, compost, and advice, and their program is to figure out something to grow and to sell. Occasionally they'll help me out with my work, but that is not the essence of the relationship. Their job is to make an attempt at profitable farming without risking any capital. And mostly they have done very well.

At this point, I continue to harvest the olives by myself, with the expectation of losing a good part of the crop. I'm not complaining about this; I enjoy working alone, and value solitude, and appreciate the occasional company of hares and crows. But this exemplifies a common problem of small farms. Lacking sufficient capital and land to compete in the realm of mechanized farming, the small-scale farmer is forced into growing labor-intensive crops, and then comes to find that the labor available is insufficient.

For apricots, the thoroughness of the harvest is even worse than for olives. I estimate my harvest (500 kg) at less than 20 percent of the crop. The crop comes into maturity very rapidly, and almost immediately it is set upon by flocks of birds who damage much of the fruit. Most of the apricots are made into apricot jam, and this is a rate-limiting step. We use a small batch process, and the fruits cook for three hours in order to get the desired thickness of texture without using added pectin. So jam making can handle less than a hundred pounds a day. As the maturity of the crop advances, the apricots start falling from the trees. Inevitably, in the course of harvesting, a lot of fruit is knocked to the ground.

The situation with figs and with blackberries is similar to that of apricots, though not quite as bad, with harvest of perhaps 40 percent of the crop. Here too, animal predation and the rate-limiting aspect of jam making come into play. Watermelons and cantaloupes have a harvest rate of less than 50 percent. Most of the loss is due to animal damage to the fruits, or uncontrolled overproduction.

You could make the case that I have too many trees of apricots and olives, and that I could do away with half of them with no loss of net harvest. This is not entirely true. There are years, usually every second year with olives, and every five years or so with apricots, when the trees produce very little fruit. It is to be able to get the fruit that I need in those years of a poor crop that I keep an excess of trees. The only way to be certain of having enough is to have too much.

For the flower crops, the rate of harvest is much better. Valuable crops like tulips, peonies, sunflowers, gerbera, and tuberose are harvested with a thoroughness well above 90 percent, and the remaining crops are above 50 percent. Our average annual production of flowers, in the form of bunches (flowers of one kind) or bouquets (mixed kinds of flowers and foliage) is 5,000 to 6,000 units.

Productivity

Productivity is a ratio, with the numerator referring to the harvest (weight, or value) and the denominator referring to some aspect of production (area, or labor input, or capital invested), generally for a period of one year. So tons per acre, and dollars per acre, and tons per hour of labor, and dollars per hour of labor, or dollars income per unit of capital invested, are all classic measures of productivity. For example, the productivity of my olive trees in the field is six to eight tons per acre, but their productivity based on harvest is five tons per acre or less. The productivity of the harvested olives, once the olive oil is extracted and bottled and sold, is $6,000 to $12,000 per acre.

While useful, the classic measures of productivity ignore the nutritional value of the crop, and always ignore the ecological value, whether it is positive or negative. For example, my olives are producing about 460 kg per acre of olive oil (based on the 50 percent harvest rate). Sugar cane produces an average of about 22,000 kg per acre of pure sugar. The energy content of the olive oil is 14,800 MJ, while the energy content of the sugar is 374,000 MJ, twenty-five times the energy production of olive oil. However, sugar is an unhealthy food while olive oil is very healthy. If we consider the production of antioxidants, a powerful health-giving component of olive oil, productivity is 7.5 kg/acre for olive oil and zero for the sugar. And if we estimate ecological impact, olive trees, which are ecologically benign and which stabilize the soil and provide habitat for wildlife, are far superior to sugar cane, which is highly destructive of the soil. So an acre of sugar cane with its exceptional productivity of energy may go farther in the short run in feeding hungry people than would olive oil; nonetheless, by other measures of productivity, olives are clearly a superior crop.

One other useful way to measure productivity of the farm is to calculate the number of people who can be fed from the farm; like the efficiency ratio, this is based entirely on energy. For a standard diet of 2,000 calories per day (730,000 calories per year), the farm produces enough food calories to continuously feed 44.6 people. Leaving aside land devoted to growing flowers or in natural habitat, this works out to about 4.5 people per acre, slightly less than grains, but higher than fruits and vegetables.

Productivity and Energy Efficiency

Energy efficiency is essentially another measure of productivity; that is, it is a ratio, of food energy created to energy invested. The aboriginal peoples of this region were hunter-gatherers, with acorns and native black walnuts the principal forage crops. The energy efficiency of gathering nuts and acorns is very high; with a heavily laden tree, an hour of labor can yield more than a month's worth of food. So the aboriginal economy may be considered to have an unusually high productivity in the sense of energy produced per input of labor energy, but a very low productivity in terms of energy produced per acre of land. The evolution of modern agriculture could be described as a history of decreasing energy efficiency in exchange for increasing per-acre productivity. It is easy to imagine situations in which maximizing per-acre productivity is no longer optimal, even in purely economic terms.

Productivity and Ghost Acres

Chris Smaje used the term "ghost acres" to refer to land outside the boundaries of the farm, from which organic matter is taken, composted, and brought to the farm. Trucking in compost distorts the apparent productivity (per acre) of the farm because the productivity of the ghost acres contributes to yield of the farm without those acres being included in the denominator. Because of this, calculations of productivity on land receiving compost of off-farm origin will be too high.

I use two main off-farm inputs. The first is gypsum (calcium sulfate) mined from mountainous gypsum deposits in Utah, 800 miles distant, and shipped to California by rail. The gypsum is used to correct the imbalance of magnesium and calcium and to lower the pH of the soil, thereby increasing yield. Because of the high magnesium content of my irrigation water, I will never reach a stage where "enough" calcium has been applied unless I cease irrigating. In the long run this is clearly an unsustainable practice.

Because only about one-third of the earth's land is arable, there is a certain logic to taking materials from the other two-thirds and using them to increase productivity of the arable land. The use of gypsum from Utah is a case in point; the detriment to the landscape and ecology of Utah from gypsum mining is likely minor, at least from a human perspective.

The compost that I have delivered each year is sourced in part (about 10 percent) from restaurant waste. When food crops are harvested from the farm and sent to market, minerals in the crop are removed from the farm; that these should be returned to the farm via restaurants and the composting facility appears to be a virtuous cycle. The remainder of the compost (90 percent) is sourced from urban green waste—grass clippings, tree prunings, and leaves from parks, athletic fields, and gardens. Removing the urban green waste depletes the soils of origin, the fertility of which is maintained by heavy application of synthetic chemical fertilizers. So bringing in certified organic compost could be seen as a roundabout way of smuggling in synthetic fertilizer. And if we look at the mineral contents of the compost, the phosphorus and potassium represent unsustainable mining practices similar to the case of gypsum. The nitrogen in the compost was processed from atmospheric nitrogen using fossil fuels; those fossil fuels could be attributed to a sort of second-order ghost acreage—land on which grew the plants that eventually became the fossil fuels, remote from the farm in both space and time. At any rate, the twenty tons of compost that I purchase each year represents one to two ghost acres of the farm, and so will amplify calculations of productivity per acre.

Productivity and the Green Revolution

Proponents of the Green Revolution sought to alleviate world hunger by applying modern science to agriculture, thereby increasing its productivity. The most commendable part of the program was the breeding of sturdy, dwarf varieties of grains and pulses to replace unruly tall varieties. Replacement of the labor of draft animals and humans with simple and dependable tractors was nearly always beneficial as well. The remainder and more insidious part of the program was about applying fossil fuels and capital to world agriculture. This included heavy use of synthetic fertilizers and pesticides, extreme mechanization of farming, and development of irrigation on previously unirrigated lands. The industrialization of agriculture made it accessible to exploitation by distant sources of capital.

The outcome, on a global scale, was for the most part socially and ecologically harmful: a flood of cheap grains from the US, Canada, Australia, and the Ukraine undermined local food economies and bankrupted small farmers, many of whom ended their lives in urban poverty; consolidation of land holdings previously owned by small farmers allowed for large-scale capital-intensive farming of commodities for export rather than food for local consumption; and heavy investments of foreign capital meant that the wealth created by agriculture flowed to the centers of capital (Europe and North America) instead of remaining in the country of origin. Productivity, in pounds per acre, did increase on lands heavily fertilized and irrigated compared to the same land unfertilized and unirrigated, and this increase in productivity is the basis on which the Green Revolution collected its accolades and Nobel prizes. Other measures of productivity, however, are negative: biodiversity per acre—decreased; jobs per acre—decreased; energy efficiency—decreased; local income from agriculture—decreased; soil carbon per acre—decreased; CO_2 emissions per acre—increased. The industrialized nations importing commodity agricultural products from countries of the tropics are able to reap the benefits while leaving the ecological and social costs in the country of origin—a modern extension of traditional imperialism.

The Green Revolution proved a powerful stimulus to the industrialization, commodification, and globalization of agriculture. This is usually presented as a benign process of plant breeding and favorable technologies, but the reality is that the Green Revolution is an amalgam of technology, government policy, finance, and relentless propaganda. The outcome is the global economic and culinary landscape in which small farms like mine must try to find a niche.

Chapter 11

Economics

There is a widespread tendency in our culture to view all human activity in financial terms—dollars acquired and dollars disbursed. At no other period in history have human values been so thoroughly monetized. I reject that view. For me, farming is not primarily about money. It has more to do with an interesting and enjoyable way of living, and with having a useful role in my community (a community that includes not only humans, but also the other organisms with whom we share the region). Nonetheless, I am forced to deal with the economics of farming as part of the reality of 21st-century life. The farm is our sole source of household income, and there have been years when an insufficiency of income has been inconvenient and stressful.

A good way to get a snapshot of the economic status of the farm is to look at the income statement that is submitted each year to the Internal Revenue Service. This is reported on Form 1040, Schedule F: Profit or Loss from Farming. An example is given in the accompanying figure, and it is worth going over it line by line, as this will illuminate many of the economic aspects of farming.

Line 2: This is the gross income from sales, both retail and wholesale, of products produced on the farm.

Line 4a: According to the USDA census of agriculture, farms with gross incomes similar to mine receive, on average, $6,500

SCHEDULE F (Form 1040)
Department of the Treasury Internal Revenue Service (99)

Profit or Loss From Farming

▶ Attach to Form 1040, Form 1040NR, Form 1041, Form 1065, or Form 1065-B.
▶ Information about Schedule F and its separate instructions is at *www.irs.gov/schedulef.*

OMB No. 1545-0074
2015
Attachment Sequence No. **14**

Name of proprietor | Social security number (SSN)

A Principal crop or activity | B Enter code from Part IV ▶ 1 1 1 3 0 0 | C Accounting method: ☒ Cash ☐ Accrual | D Employer ID number (EIN), (see instr)

E Did you "materially participate" in the operation of this business during 2015? If "No," see instructions for limit on passive losses ☒ Yes ☐ No
F Did you make any payments in 2015 that would require you to file Form(s) 1099 (see instructions)? ☐ Yes ☒ No
G If "Yes," did you or will you file required Forms 1099? ☐ Yes ☐ No

Part I Farm Income—Cash Method. Complete Parts I and II (Accrual method. Complete Parts II and III, and Part I, line 9.)

1a	Sales of livestock and other resale items (see instructions)	1a				
b	Cost or other basis of livestock or other items reported on line 1a	1b				
c	Subtract line 1b from line 1a				1c	
2	Sales of livestock, produce, grains, and other products you raised				2	99,340
3a	Cooperative distributions (Form(s) 1099-PATR)	3a		3b Taxable amount	3b	0
4a	Agricultural program payments (see instructions)	4a		4b Taxable amount	4b	0
5a	Commodity Credit Corporation (CCC) loans reported under election				5a	0
b	CCC loans forfeited	5b		5c Taxable amount	5c	0
6	Crop insurance proceeds and federal crop disaster payments (see instructions)					
a	Amount received in 2015	6a		6b Taxable amount	6b	0
c	If election to defer to 2016 is attached, check here ▶ ☐			6d Amount deferred from 2014	6d	0
7	Custom hire (machine work) income				7	2,000
8	Other income, including federal and state gasoline or fuel tax credit or refund (see instructions)				8	0
9	**Gross income.** Add amounts in the right column (lines 1c, 2, 3b, 4b, 5a, 5c, 6b, 6d, 7, and 8). If you use the accrual method, enter the amount from Part III, line 50 (see instructions) ▶				9	101,340

Part II Farm Expenses—Cash and Accrual Method. Do not include personal or living expenses (see instructions).

Line	Description	Box	Amount	Line	Description	Box	Amount
10	Car and truck expenses (see instructions). Also attach Form 4562	10	4,211	23	Pension and profit-sharing plans	23	0
11	Chemicals	11	0	24	Rent or lease (see instructions):		
12	Conservation expenses (see instructions)	12	0	a	Vehicles, machinery, equipment	24a	135
13	Custom hire (machine work)	13	750	b	Other (land, animals, etc.)	24b	0
14	Depreciation and section 179 expense (see instructions)	14	13,266	25	Repairs and maintenance	25	2,655
15	Employee benefit programs other than on line 23	15	0	26	Seeds and plants	26	11,444
16	Feed	16	0	27	Storage and warehousing	27	0
17	Fertilizers and lime	17	808	28	Supplies	28	25,699
18	Freight and trucking	18	98	29	Taxes	29	2,115
19	Gasoline, fuel, and oil	19	624	30	Utilities	30	3,888
20	Insurance (other than health)	20	175	31	Veterinary, breeding, and medicine	31	0
21	Interest:			32	Other expenses (specify):		
a	Mortgage (paid to banks, etc.)	21a	0	a	FARMERS MKT. FEES	32a	5,004
b	Other	21b	0	b	LICENSES & PERMITS	32b	855
22	Labor hired (less employment credits)	22	0	c	ORGANIC CERTIFICATION	32c	725
				d		32d	
				e		32e	
				f		32f	

Line	Description	Line	Amount
33	**Total expenses.** Add lines 10 through 32f. If line 32f is negative, see instructions ▶	33	73,452
34	**Net farm profit or (loss).** Subtract line 33 from line 9. If a profit, stop here and see instructions for where to report. If a loss, complete lines 35 and 36.	34	27,888

35 Did you receive an applicable subsidy in 2015? (see instructions) ☐ Yes ☒ No
36 Check the box that describes your investment in this activity and see instructions for where to report your loss.
a ☒ All investment is at risk. b ☐ Some investment is not at risk.

For Paperwork Reduction Act Notice, see the separate instructions. Cat. No. 11346H **Schedule F (Form 1040) 2015**

IRS Form 1040, Schedule F: "Profit or Loss from Farming."

each year in direct subsidies from the federal government. The subsidies are not distributed equally. They are as much political as economic, and they are heavily slanted toward commodity crops. Specialty crops such as those that I grow are not subsidized.

I am eligible for a partial refund of the $725 fee that I pay for organic certification through a USDA organic program grant. In the past I sometimes took advantage of this, but I no longer do so. The organic certification industry is a racket—another bureaucracy carried on the backs of farmers, and so some relief from those expenses seems fair. But this strikes me as an inappropriate use of funds from the public treasury, and I'm reluctant to take part in it. I know farmers who make a deep study of federal and state programs, and grab every dollar that might be available to them. My practice is the other way; I have not taken advantage of subsidies for solar power installation, conservation practices, development of value-added projects, efficient irrigation, and others of my farm activities that would be eligible for tax credits or grants. Similarly, there have been years of low income when we were eligible for food stamps and other sorts of public support, but we have never enrolled in those programs. My convictions on this topic are still somewhat fluid, but it seems to me that one should do whatever it is that is the right thing to do, and to accept public monies for it is irrelevant and dishonorable. (I also must confess to a certain element of laziness about completing all the paperwork that claiming public funds requires.)

Line 6: I do not use crop insurance—I just take my losses as losses.

Line 7: This income represents custom milling of 5 tons of olives for friends at $400 per ton.

Line 10: We are only 8 miles from our farmers market and from a supermarket where we sell our products, so our annual mileage for farm business is small. We know other farmers who log 500 miles a week or more trucking their produce to markets in the San Francisco Bay area. The dollar cost of trucking (as well as its time cost and ecological cost) is quite high in those circumstances.

Line 11: Chemicals. On conventional farms, this is a significant item of expense.

Line 13: This charge is for a contractor who brought in a chipping machine to chip all the prunings from the orchards.

Line 14: Depreciation. This includes machinery, tools, and buildings. It is somewhat fictional in the sense that a well-built and well-maintained building is actually increasing in value, not decreasing.

Line 16: Because we do not sell the eggs, but give them away, we cannot deduct the cost of feed for the chickens. Nor can we deduct the cost of feed for cats, even when they are kept primarily for rodent control.

Line 17: Compost and gypsum.

Line 18: This covers a repair part for the olive mill sent from Italy.

Line 19: Fuel for tractors and tools. Fuel for vehicles is covered in line 10.

Line 21: I have no debts, so I pay no interest. I understand that credit is a tool, but it is one that we do not use.

Line 24: Rental of a trailer and of a trenching machine.

Line 25: This includes expensive remedial work on a water well. Every year it seems that some major repair will be required. I don't know in advance what it will be, but I know that something will break in a way that is costly to repair.

Line 26: We plant a lot of bulbs every year (tulips, iris, ranunculus, lilies, etc.), and also continue to add trees—primarily citrus and olives. Some seeds, such as hybrid flower seeds, may be very expensive.

Line 28: The majority of this is packaging: bottles, caps, capsules, jars, lids, labels, bags, and boxes. We have not found suitable domestic glass, and so the bottles for olive oil are imported from Italy. Customers return about 40 percent of the glass, which we are able to clean, sterilize, and re-use. Also included in this line are a hundred dozen other little things—work gloves, paper towels, mouse traps, saw chains, sugar for jam, etc.—that add up in the course of a year.

Line 29: This includes that portion of the property tax attributable to the farming business, as well as sales tax collected on taxable items (flowers) sold in the farmers market. When we sell a bunch of flowers for five dollars, the true price is $4.63 plus 37 cents tax. That tax must be remitted to the state.

Line 30: Electricity for pumping water, running the cooler, running the olive mill, and charging the electric truck. Lighting is all LED lights, and does not use much power.

Line 32a: The farmers market fees are based on a percentage of gross sales. They pay for management, trash pickup, insurance, utilities, and the like. The market is very well run, but expensively so.

Line 34: The net income (incorrectly called "profit" on the form) is only about one-quarter of the gross revenue. This is typical of farming operations whether they are large or small.

Beyond Schedule F

In table 9.1 (page 84), I summarized the hours of labor that we (two adults) expend annually on the farm at 6,073 (3,036 hours each). In the US, the current annual labor of a full-time job is 1,900 hours: 52 weeks of 40 hours minus vacations, holidays, and paid leave. So our work on the farm adds up to somewhat more than three full-time jobs done by two people. I generally work about 350 to 360 days each year. Dividing the net revenue of the farm by the hours worked gives an hourly wage of $4.59, at a time when the minimum wage in California is $11.00 per hour, scheduled to rise to $15.00 in the near future. Enforcement of a minimum wage does not apply to the self-employed, and an income below minimum wage is typical for small-scale farmers. So, based on Schedule F, the income from the farm pays a substandard wage, while the profit is zero and the return on investment is zero.

Although the amount of hours worked might seem excessive, by historical standards it is not. There are 8,766 hours in a year, which leaves each of us 5,730 hours outside of work for other uses. And it's worth noting that this is not some soul-crushing office job or factory job; this is pleasant, interesting, autonomous, meaningful work carried out under the open sky.

The hourly wage calculated from Schedule F is misleading, and it is important to consider the inaccuracies that Schedule F engenders. For example, vehicle expenses are charged at a per-mile rate, currently about 56 cents. But I purchased my vehicles used at a good price, I maintain them myself, and I drive conservatively, so that the true cost of operation is closer to 30 cents per mile. The $4,211 vehicle expense listed on line 10 is in reality only about $2,200. Similarly, the depreciation of farm buildings is considered an annual cost, when in reality the buildings are increasing in value. And since the buildings were paid for at the time they were built, the annual depreciation expense is hypothetical, and does not actually take any money out of our pockets in the current year.

A second point overlooked by Schedule F is that we grow a lot of our own food as part of the farming operation. In addition, we barter produce and olive oil for wine, avocados, strawberries, veterinary care, and a few other things. In theory one is supposed to report the value of barter as income, but on this scale—mutual gifts among friends—the rule does not apply. These activities represent income with a value of about $5,000 per year that does not show up on Schedule F because it is not monetized, and because the government has not yet tried to tax people for growing their own food.

Most important, though, is that much of our labor increases the value of the farm, and hence our wealth, without ever manifesting as income. For example, if I plant an orchard, and dig trenches for irrigation, and lay pipe, and plant each tree, and drive in a stake and tie off each tree, and prune the trees, and keep after the weeds, and trap gophers, all of this labor to establish the orchard is not rewarded with any income. However, it has increased the value of the farm by somewhere between $3,000 and $6,000 per acre planted.

As it turns out, erecting buildings is the most profitable use of my time on the farm. When I do farm labor I'm replacing an $11 dollar per hour farm worker, but when I'm building I'm replacing a $50 per hour building contractor. I built the building that houses the olive mill and jam kitchen for a total cost of $17,000 ($2,400 was the building permit) using mostly recycled materials. This took me a year, fitting the labor in between my regular farming work. A contractor would have charged at least $100,000 to build that building. So I came out ahead by $83,000. This does not show up as income, but it constitutes an increase in our wealth.

Our total cash outlay for our farm land, buildings, trees, and infrastructure, not adjusted for inflation, is approximately $200,000. The current market value of our farm, if we were to sell it, is above $1,500,000. While nearly all of that increase represents inflation, some of it—perhaps $250,000—represents the value of labor applied to the farm that was never compensated as income. In addition, the residual value of tools and machinery that have already been depreciated is about $80,000. We can conclude that our annual income is not as bad as it appears on Schedule F; we probably make minimum wage.

Risk Management

Part of the economic management of a business involves dealing with issues of risk. Our approach has been one that a wise business advisor would not approve of, but it has worked out for us so far. We carry minimal insurance: the mandatory insurance on our vehicles and fire insurance on our house. When classes of school children come out for a farm tour, I am taking on a certain amount of risk that one of the children might be injured in some way and I would be found responsible. Some farmers carry expensive farm insurance policies and require signed liability waivers from visitors, but I just trust in people's good sense and good will, and so far that has worked out.

I do not carry crop insurance, although it is cheaply available, subsidized by the federal government. My approach has been rather to grow a wide diversity of crops so that failure of a few is mitigated by the success of the others. This approach has served us adequately.

The greatest risk to our operation is injury or illness of either of us. An incapacitating injury would mean the end of the farming operation. There was a year when I was ill for a long period, and during that time the productivity of the farm declined alarmingly. But other than that we have been healthy and able to work. In a large corporate/industrial farming operation, this risk is alleviated by having a hired staff of farm managers. If one of the managers is injured or ill or retires, he can be replaced by a new hire with no discontinuity in the overall operation.

Income History

At the outset we had years of negative income from farming. In those early years I had a day job. After a few years I was able to reduce my off-farm employment to part-time, and then, as we gained financial momentum, I gave up off-farm work in year 11 in order to farm full-time (time and a half, actually). The steadily rising income represents increasing intensification and mechanization, as well as increasing skills and knowledge that let us improve the efficiency of the farm.

Around years 16 to 18 there is a subtle increase in the growth rate of our income, which represents a shift from selling only fresh produce to selling value-added products (jam, olive oil, soap) as well. It is the nature of our food economy that most of the profit is made by processors and distributors, not by farmers. And so, by becoming a processor, we are able to capture some of that downstream profit. The other obvious advantage of processing our harvest is that it converts a highly perishable product into a stable product that can be sold throughout the year. In the early years we had no sales in the month of December; now, December is our best month for sales.

Options to Increase Farm Income

Our operation produces a low current income. I have considered four principal ways in which we might remedy that; probably there are others that I haven't thought of. While increased income would have been helpful in the past, it is no longer especially urgent. Our children have finished their university studies and are on their own; we have no debts; and our current requirements for income are very modest. Nonetheless, it is instructive to consider ways in which farm income might be increased.

The first is to increase our efficiency and productivity. We could mechanize our olive harvesting, but our scale is too small and our remaining years of farming too few to justify that expense. We could revamp jam making from a small-batch artisanal process to a larger-scale industrial process, but that would change the nature of the product in a way that doesn't work for us.

A more promising route would be to develop a line of table olives to be sold in jars or in bulk. A box of freshly harvested olives is worth five times as much made into table olives as it is processed for olive oil, and there is a strong, unfulfilled demand for table olives. We have been experimenting for several years with making table olives, but we have not yet learned how to predictably make a successful batch. The harvest of green olives for table olives takes place before harvest begins for olive oil, so there is no scheduling conflict. This is a work in progress, one that might eventually boost our income. Producing table olives for sale requires a costly cannery license with related certificates and permits, so the undertaking is economically feasible only if we start off at a fairly large scale; producing only 500 jars of table olives could not be made to cover the permitting costs.

The third option we have to increase farm revenue is to add some component of agritourism. This is a boon to many small farms around the country, including some in our neighborhood. The farm can be rented out for events, or farm-to-fork meals, or harvest festivals, or a bed and breakfast quarters can be added. Several farms in our area specialize in weddings, which are far more lucrative than farming. There is always a danger with this that the farm loses its authenticity as a farm and instead becomes a sort of stage set. (We live in the era of virtual reality and the fake experience; a stage-set farm is in keeping with the spirit of the times.)

Our farm would be a natural agritourism site. Many people are interested to visit a place with five acres of flowers, and in this dry country a happily flowing creek is an irresistible attraction. But we have no plans to pursue agritourism. It does not fit our personalities or our interests, and we prefer to dedicate ourselves to operating the farm.

The last option, which is the most obvious way to increase our income, is to raise our prices. Unlike fungible commodities where the price is set nationally in futures trading markets, the artisanal products of our farm can be priced however we think appropriate. There is widespread agreement among small-scale farmers that you should set your prices high. One of the farming periodicals advises that if you're not getting complaints about your prices, then they are too low. But pricing is not so simple, because in addition to its economic aspects it is also bound up with issues of social justice and wealth inequality.

Consider the situation of selling olive oil at the farmers market. Here comes a shabbily dressed young woman with three small children, and she carries a fistful of wooden EBT tokens (the market's version of food stamps). Clearly, this customer is struggling, and I'm happy to sell her a bottle of olive oil for $5, or even just give it to her. The next customer is an arrogant and imperious woman laden with gold jewelry who probably believes, incorrectly, that her wealth, acquired through no virtue of her own, is evidence of her superiority to the rest of us. I would want to charge her $40 for a bottle of oil. Obviously, this kind of variable pricing cannot work, and in an unjust society there is no such thing as a just price.

Many of the artisanal olive oil producers price their oil at $25 to $45 per half-liter bottle. We charge a retail price of $12 for a half-liter bottle, and most of our oil is sold in 3-liter jugs with a unit retail price of only $8 per half-liter. Why the difference in price? Many of the producers have a large capital investment, buying expensive land and hiring laborers, builders, millers, accountants, graphic designers, and a host of other workers. Even with high prices they will operate most years at a loss. Our much lower overhead allows us to sell at a lower price. But there is also a difference of philosophy. They believe that the goal of business is to make as much money as possible (the conventional business school view), whereas our goal is to be useful and reliable members of our community. And since our community includes a lot of poor people, we keep our prices low, and we donate olive oil and produce to organizations that feed indigent people. Those who set their prices high serve only the wealthiest stratum of the community. It also strikes me that a high price implies excessive self-esteem on the part of the vendor; for me, this is another obstacle to raising the price.

This is a confusing subject, and in some irrational way it sets the upper limits on our prices. One element of this is that, over time, each person develops a sense of the exchange value of labor and goods. Because we both started out working as children, and for many years had low-paying jobs, we tend to put a low exchange value on our labor, an idea reinforced by my years of living in peasant and aboriginal regions of the tropics. Our prices are low, and we get complaints from other farmers that we are undermining the market. And yet, we are unwilling to be farming only on behalf of rich people. And if that impoverished woman with the three small children asked the price of olive oil, and I had to tell her it was $35, I would be deeply ashamed.

An underlying problem with price is that industrial commodity foods that make up the great majority of food in this country are seriously underpriced. This reflects direct subsidies (cash payments, price supports, crop insurance), semi-direct subsidies (infrastructure such as irrigation districts, publicly employed farm advisors) and hidden subsidies. The latter include the usual externalities: bad foreign policy, an expensive military, underpriced energy, rigged markets, damage to public health by pesticides, inadequate compensation of workers, and environmental degradation of many kinds. The small-scale farmer who farms in an ecologically and socially sound way is denied these subsidies, and thereby suffers a disadvantage. In the US, food, whether industrial or artisanal, should fetch a higher price than it does, and while I encourage other small-scale farmers to seek higher compensation, I am not comfortable with that for my own business. I have been told that I am a good farmer and a poor businessman—probably true.

Enough Income

Late one evening there was a knocking on our door, and I answered to find a man from China who had heard of our operation and wanted to talk to me. I invited him in and we talked for a couple of hours, and at the end of our conversation he offered me a job. The proposition was to plant 5,000 acres of olive trees in far western China where he had an option on suitable land, to set up a an olive oil mill, and establish an olive oil company. "Name your salary," he said. "There's plenty of money in this project—any amount is okay." When I declined his offer, he said, "I don't understand Americans. No one seems to have an ambition to be rich. In China, everyone wants to be rich. It's all we think about. It's all we talk about."

I explained my views on this, and I drew a graph on the back of an envelope. One axis was money, and the other was happiness, and I pointed out to him that the graph has an inflection point beyond which further money in any amount does not increase happiness. He studied my drawing, but seemed unconvinced.

Dianne and I have never been motivated to be rich in terms of money. We live in a beautiful place, we have many friends, we're healthy, we have meaningful work, and we have wholesome food to eat and good local wine to drink—what would we want with more money? Our aim has always been to stay within what Ivan Illich called "the narrow range that separates enough from too much." (*Energy and Equity*, 1974). Our current income keeps us in that range. It seems adequate, and we're not motivated to increase it.

Chapter 12

The Social Context

It might seem that the farmer out plowing his field, or gathering fruit, or hoeing weeds, leads an autonomous life of self-determination and independence. While this may be true to a greater degree than the situation of a state or corporate employee, the farmer nonetheless finds his plans and activities circumscribed by a variety of social contacts—economic, informational, or regulatory, and ranging from the informal to the rigidly defined.

Economic Contacts

We depend on dozens of suppliers for goods and services. Some of these come out to the farm for such things as refrigeration repair, pump repair, or propane delivery. Others operate stores and warehouses where we buy things that we need, and quite a bit of our business is carried out by phone or internet with people who we have never met. We deal with eight suppliers of plant materials and seven suppliers of packaging materials, as well as dozens of businesses providing us with machine parts, tillage tools, cleaning supplies, and thirty dozen other sorts of things that we need around the farm, ranging from chemical analysis of our well water to copper jam pots. At least six courier services regularly drop off goods at the farm.

We pay our bills promptly, and so nearly all of these relationships are good ones. We recognize that many of our business associates have religious and political views that are far removed from our own, but that's not the subject of our association. If a man can fix the flat tire on my tractor, it doesn't matter to me that he's a Jehovah's Witness and a member of the Tea Party, nor does he care if I am an atheist and a Green Party voter. Because these business associations are mutually beneficial, everyone involved is motivated to be tolerant. If we look at historically pluralistic societies, the mutual dependence and benefit of business enforced cultural tolerance, and that is still true.

On the sales side of the equation, nearly all of our sales are direct to the customers through the farmers market. Our business plan, insofar as we have one, is of a sort that might be found in 18th-century Italy or Spain. That is, we produce on a small scale and sell our goods in the local town where we are well known. We have no need to advertise, or to brag on the internet, or to vie for gold medals at the state fair; our local reputation insures that everything gets sold. Our relationships to our customers, most of whom we know, are both economic and social. Amazingly, Dianne remembers the names and situations of close to a thousand people, which is highly beneficial to our business; I struggle to remember the names of a few dozen of my customers.

The Informational Matrix

Farming is a moderately information-dense undertaking, and it is important to keep current. In this I depend quite a bit on the University of California. The Cooperative Extension portion of the university puts on short courses and field days that are very useful. Other pertinent information is readily available from university publications or by telephoning an extension agent. I also rely on the university's library. Five or six times a year I will spend half a day at the library at UC Davis reading agricultural journals. I particularly seek out journals from places with a climate similar to ours—South Africa, Australia, Chile, and the Mediterranean—to see what farmers in those places are doing.

Farm equipment expositions, held each year in the wintertime, are another good avenue for finding out what's new. Not only farm equipment is featured: vendors of all manner of services, insurance, farm publications, solar conversions, unorthodox fertilizers, and many other marvels are there promoting their wares.

There are two relevant farm conferences each year: the Small Farm Conference and the Eco-Farm Conference. These tend to be repetitive from one year to the next, but I attend both of them once every eight or ten years.

One organization that we do not patronize is the California Farm Bureau. While it has a valuable informational component, it seems to be primarily a lobbying organization on behalf of agribusiness. Its values are strongly pro-business and anti-environment, and it is consistently anti-government (except, of course, for farm subsidies and water subsidies). The Farm Bureau's suggestions on how to vote in state and federal elections are nearly always the opposite of how I vote. But even if their views were closer to mine, I believe that cash-intensive lobbying organizations are destructive of democracy, and should not be supported.

For me, the main source of information is other farmers. There are between twenty and forty farmers at our farmers market each week, and we walk about the market and greet one another, complain about the weather, and discuss the issues of farming that are of concern at the moment. This is by far the most lively, diverse, and up-to-date source of information for all of us. Although we are competitors, at the same time we are colleagues and friends.

Finally, the farm itself almost always provides the information that I need. I notice that young olive trees are being defoliated by a pest that chews the margins of the leaves. I could phone the extension agent, or bring a sample of leaves to her office, or check at the UC Davis Olive Center, but instead what I do is to go out late at night with a flashlight and discover that the culprit is a small brown weevil. I collect a sample of the weevil in a plastic bag that I put in the freezer for a few hours to kill it, and then at my leisure I can consult my books to discover the weevil's name (*Otiorhynchus cribricollis*) and habits. My treatment for this is to douse the trees with powdered diatomaceous earth, which is more or less effective.

The flow of information is not unidirectional. I host tours of many groups through the year, ranging from school children to college students to professional groups, such as the master olive millers. Students from the California Farm Academy spend a half day getting a close look at the operation. Because the farm is adjacent to the campus of UC Davis, we get many foreign visitors who are interested in small-scale agriculture. Other people—mostly would-be farmers—find us somehow. We get about twenty visitors annually in addition to the many groups that visit. Some farmers charge a fee for tours, but I consider it to be community service and do not charge. Probably I learn as much from the visitors as they do from me, especially the visitors from other countries.

Regulatory Agencies

The farmer of 120 years ago could build his house and barn, raise his crops, butcher his hogs, make and sell jam from his fruit, brew his beer, burn his trash, drain his fields, sell his produce, pee against the side of the barn, and in general live however he saw fit without any outside supervision. Some of the farmers did a good job, and some made a mess of it, and each lived with the consequences of his own decisions and actions. This is no longer the case. Nearly every aspect of farming is subject to regulation. A list of the government agencies with which I have farm-related business is given in the sidebar on page 122.

I won't run down the full list of regulatory agencies, but I will illustrate my interactions with two of them. The California Department of Public Health issues the food processor registration that constitutes the permit for the production of jam, olive oil, and skin care products. The fee is $548 per year. Every second or third year I get a random, unscheduled inspection to assess sanitation and compliance of my facility. The inspection lasts about fifteen minutes. The inspectors are knowledgeable and reasonable in pursuing their principal concern, which is food safety. I think that the program is appropriate, and I have no complaints with it except for the price—$1,096 for a fifteen-minute inspection.

My membership in the Dixon/Solano RCD Water Quality Coalition is mandated by state law. The mission of this agency is to ensure the safety of groundwater and to protect it from contamination by nitrates and pesticides. The annual fee is reasonable ($3 per irrigated acre), but membership has other requirements that are time-consuming. These include a requirement to participate in at least one short course per year and a requirement to file a nitrogen management plan. Because my farm is designated as highly vulnerable to nitrate leaching, additional requirements come into play, including a requirement that the nitrogen plan be signed by a certified plan preparer. Given that this obligation recurs annually, I found the easiest path for me was to take the course and pass the exams to become a certified nitrogen planner myself. The goals of this agency are worthy, but are directed entirely at conventional farms, and are not really pertinent to my operation, since I use neither pesticides nor synthetic nitrogen fertilizers. There is no way to opt out.

The other agencies on the list vary from some that are benign to others that collect fees, sometimes unreasonably large, simply because they can. There is a recurrent pattern to the growth of bureaucracies. In the early stages they are nimble and mission-driven, but as the bureaucracy grows, the bureaucrats lose sight of their mission and become devoted to protecting their own turf and perquisites, while internal administration comes to dominate the budget. (This illustrates what Joseph Tainter called "the declining marginal utility of increasing complexity" in *The Collapse of Complex Societies*, 1988.) The building department

Government Agencies Regulating Farm Activities

- Solano County Agricultural Commissioner
- Solano County Assessor
- Dixon/Solano RCD Water Quality Coalition
- Solano County Planning Department
- Solano County Building Department
- Solano County Department of Resource Management
- Dixon Fire District
- Yolo Certified Organic Agriculture
- Yolo County Sealer of Weights and Measures
- California Department of Food & Agriculture (CDFA): Organic Registration
- CDFA—Farmers Market Inspection Program
- CDFA—Citrus crop report
- California Department of Motor Vehicles
- California Department of Public Health
- California Board of Equalization
- California Franchise Tax Board
- California Labor and Workforce Development Agency
- Sacramento Valley Water Resources Board
- California Air Resources Board (CARB)
- USDA Horticultural Survey
- USDA Census of Agriculture
- USDA Agricultural Resource Management Survey
- US Food and Drug Administration
- Internal Revenue Service
- United States Customs

provides an example. During a building boom they hired a lot of extra staff, and when the pace of building declined they didn't want to lay anyone off, so they compensated for overstaffing by greatly increasing the complexity of the permitting process and the fees charged. To erect a tractor shed on my farm would require multiple sets of engineered plans; approval by seven different sub-departments, each with its own fees; a check of the full property infrastructure against recent aerial photographs and a history of previous permits to verify that no unpermitted construction had taken place; payment of a school impact fee (although there is no impact on schools); approval by the fire marshal, including inspection of my driveway to verify that two fire trucks could pass each other without running off the gravel or being scratched by a bush; and extra copies for the assessor's office so that they can raise the taxes as soon as possible. If it is a metal building, I am required to hire an outside certified bolt inspector to measure torque on the bolts in order to verify that they are adequately tightened. All of this requires several months of review and payment of fees that can amount to more than half the cost of the structure. Why shouldn't a farmer be entitled to build a humble structure like a tractor shed without all this rigamarole? And having built such a shed, why should he have to pay $300 each year in additional property tax for the rest of his life? Common sense has entirely disappeared in this process. For me, the upshot of this is that my tractors sit out in the weather rather than in a shed.

The most troublesome agency is the California Labor and Workforce Development Agency, which establishes and enforces rules for labor. The effect of their policies is that the mandated cost of hiring labor—wages plus overhead—is so high that I cannot afford it. Starting when I was eight years old, I always had jobs, some of them on farms. There was no regulation, no Social Security or taxes withheld, no insurance, and no reporting. A farmer needed someone to cut apricots in the fruit-drying yard. The pay was 38 cents for a well-filled tray, three feet by six feet. The pay was okay with me; I was happy to have the opportunity to work, and the farmer needed the help. So I went to work, and was paid in cash. I liked working—it made me feel useful and important, and having my own money gave me a whiff of independence. What business is it of the state to prohibit such an arrangement?

I would like to hire a high school student to help me out in the summer. I could afford to pay $200 a week. This would let the student learn some skills, and get some exercise, and earn a useful sum of money, and develop a sense of self-worth, and I would get some help that I need, and perhaps a chance to share some of my hard-earned knowledge. But the legal minimum cost (wages plus taxes plus unemployment insurance plus workers compensation insurance, etc.) would be over $680 a week (quite a bit more than I make myself), which I cannot afford. And so instead of having a job, the student spends his summer moping about staring at his phone, and I end up working harder than I want to.

The proponents of raising the minimum wage to $15 per hour make the case that the current minimum wage ($11) is insufficient to support a family of four, which is true. But not everyone wanting a job is supporting a family, and especially young people are denied the opportunity to work when the minimum is set too high. High minimum wage is also a strong incentive for automation and outsourcing, further eliminating jobs.

When I first planted olive trees, I understood that hand-harvesting the fruit would be labor-intensive and time-consuming. I had a vision that I would assemble a party of friends, most of them urban dwellers and office workers, and we would jointly harvest the fruit in a convivial way as is done all around the Mediterranean, and share a festive meal at the day's end. Gathering one's food is a fundamental instinct that too often is suppressed, and harvesting olives and going home with a few bottles of fresh oil at the end of the day could be profoundly satisfying. However, as it turns out, volunteer farm labor is illegal in California. The farmer must supply various kinds of insurance, minimum wage, tax sharing, and all the other obligations, even for his own relatives if they don't reside in his household.

When laws are ridiculous, civil disobedience is appropriate, and for a few years I followed that vision of a community harvest, inviting a crowd of friends to help with the olives. But I eventually gave that up. Partly this was because most of the volunteers were amazingly ineffective in their labors. Harvesting fruit is skilled labor, and takes a knack, and that knack seems to be uncommon. But another reason for abandoning the community harvest is that injury while working, even as a volunteer, is excluded from treatment by most health insurers, and the employer is held responsible instead. Clumsy computer programmers on ladders in my orchard seemed an excessive liability. So in the end, I harvest the olives myself, with a bit of help from my family.

The Shifting Nature of Laws

It used to be that the legislature would establish a law prohibiting some activity harmful to society, and if you were accused of that activity and found guilty, you would be punished accordingly. The premise was 'presumed innocent until proven guilty.' But there has been a shift in this concept toward its opposite: presumed guilty until proven innocent. Let me offer an example.

In California, a farmer can sell only produce from his own farm at a certified farmers market. If the law were construed in the traditional way, it would be assumed that all farmers were in compliance, but if one were accused of reselling purchased produce, an investigation would be made, and if the accused were found guilty he would be excluded from the markets and made to pay a fine. But instead of this traditional mechanism, the law is enforced the other way around. Each year I am required to submit to the Agricultural Commissioner a list of every crop I plan to grow in the coming year, the exact area of it, the projected amount of harvest, and the exact dates of harvest. The commissioner then sends an inspector to my farm at various times, with a charge to me of $90 per hour, to verify that I'm growing what I said I would grow. And if I decide at the spur of the moment to grow a crop that was not on my list, and if a CDFA inspector comes by the farmers market and notices that I'm selling a product that is not on my certificate, then I am considered to be guilty of an infraction and am fined, even though I am selling produce that I grew myself.

This style of regulation is increasingly common. If you wish to use the word "organic," it is not sufficient simply to farm organically and then say so. You must instead pay substantial fees to outside bureaucracies that send an auditor to inspect your farm, examine your receipts and your checkbook, snoop around in your barn, and generally assume your guilt while you are forced to try to prove your adherence to the organic rules.

Looming on the horizon is another enormous new bureaucracy within the Food and Drug Administration created to enforce the Food Safety Modernization Act. This requires of farmers that they have a food safety plan prepared by an authorized plan preparer, with enforcement requiring third-party auditors, inspections, and meticulous record keeping. Under this act, each time I take a load of watermelons to the farmers market, I must record in a log that I have cleaned my van before loading it. I don't actually have to clean the van, I just have to write those words on a scrap of paper to satisfy an inspector. Written into the law is that a reinspection is to be billed to the farmer at a rate of $221 per hour—it is clear that the bureaucrats who set that fee have absolutely no concept of what small-scale farming is about.

It does not require an evil agenda for a government to oppress its citizens; good intentions poorly executed suffice.

Some of the above sounds like the rant of an angry libertarian. Actually, I believe very much in the necessity of regulations for a well-ordered society. The tools available for an individual or a corporation to pursue mayhem and to harm society and the environment are more powerful than they have ever been, and so the need for regulation is correspondingly greater. I welcome regulations that are wisely conceived, carefully written, and thoughtfully enforced. And even regulations that are poorly crafted and perversely enforced, constituting a nuisance and a burden, nonetheless help to protect us from the follies of the ignorant and the unscrupulous.

CHAPTER 13

The Farmers

In the mid-19th century, more than half of Americans were engaged in farming, a profession that had a reputation for leading one to exhaustion, poverty, and despair. The technologies of the era dictated brutal hard labor, social isolation, and helplessness when confronted by perverse weather or infestation of crops with pests or disease. And the social and economic structure of the times enabled exploitation of the farmers by the railroads, the merchants, and the banks. But beyond those burdens is the situation that many of the people engaged in farming were not suited to it.

Being a good farmer is like being a good musician—it takes a knack. Ably strumming a guitar in your living room does not mean you can become a professional rock and roll musician, which is a far more stringent calling. And similarly, growing a productive garden in your backyard does not mean that you're suited to earn your living by farming, which requires both aptitude and a particular personality. Possibly only one person in a dozen, or one in twenty, is temperamentally suited to farming. Which is why when half the population is farming, most will do a poor job of it, and be unhappy in the process.

Three Traits of Successful Farmers

Good farmers are information-seekers. When I go to a cooperative extension field day, or take a short course, or attend a trade show, I always see the same people. Persistently seeking out useful information is a requisite for success. An old farmer may sit through a program for which 98 percent of the material is already familiar to him, but the 2 percent that is new might make it worthwhile. One of the best sources of useful information is to visit other farms and see what's going on. Picking up one good idea, or one new way to rig up equipment, or a new way of trellising a crop, can justify the day's outing. I rely heavily on reading—mostly books and journals, but increasingly the internet—to keep up to date.

A second trait of successful farmers is a tendency toward orderliness. You want to see tools well maintained and in their proper places, the stock of seeds protected from weather and rodents, flammable brush cleared away from the buildings, drainage channels kept clean in anticipation of heavy rain, and numerous other dimensions of an orderly farm and an orderly mind. If you drive by a farm where half-finished projects are strewn about, the animals have gotten loose from their enclosures, tools are rusting in the weather, and a mountain of trash intended some day to be taken to the dump is home to six dozen rats, you can conclude that this farm is a failure, even if the farmer himself has not yet realized it. On the other hand, an excess of orderliness can shade off into obsessive-compulsive disorder. The overly orderly farmer uses too many herbicides, mows too often, and is intolerant of native vegetation that might contribute to the ecological balance of the farm. The good farmer keeps to a middle path.

The third trait that predicts success is speed at work. If you send ten workers out to harvest oranges, and one worker picks 50 percent more than anyone else, that fast worker is one who can succeed at farming. When I invited urban friends to help harvest olives, I knew that some would be faster and some would be slower, but I didn't anticipate the magnitude of the difference. I thought it would be about two to one, but it was closer to twenty to one. The fastest picked 400 pounds of olives in a day while the slowest had hardly managed 25 pounds. Working by myself, early in the season when the olives are green, I harvest about 700 pounds per day, and as the fruit ripens and is more easily harvested, this rises to as much as 1,500 pounds in a day. I am a 99th percentile worker with respect to speed—not thoroughness, not accuracy, just speed. I use a lot of oxygen, and I set a fast pace and keep it up for ten hours. And this is not only for harvesting fruit but for transplanting seedlings or spreading compost, or shingling a barn.

Paradoxically, although my work is fast, it is unhurried. A friend of mine hires half a dozen women to harvest blackberries on his farm. There was one who had twice the productivity of the others. He watched her work to see how she achieved this; she worked at a comfortable pace, like the other workers, and yet, at the end of an hour, she had twice as many blackberries as they did. So that's the skill one seeks: fast but unhurried.

The Required Skills

The phrase "A jack of all trades and master of none" is usually applied scornfully. It suggests someone inconstant, a dabbler and drifter, one without focus or staying power. We live in a society of specialization, and you have to master only one trade (gall bladder surgery, corporate taxes, pro hockey) to have all the world's good things flow toward you. But on a small farm, to be a jack of all trades is a necessity and a virtue.

The successful small-scale farmer needs to be a welder, a diesel mechanic, a plumber, a carpenter, a roofer, a labor contractor, an accountant, an agronomist, a veterinarian, a pomologist, a pathologist, a chemist, a soil scientist, an electrician, a concrete finisher, an irrigator, a salesman, an arborist, a truck driver, a beekeeper, and a few dozen others. Of course, he can hire people to do these various things, but hiring specialists is expensive, and the economic reality is that at most he can hire two or three. Hire four and his income drops to break-even. Hire five or more and he operates at a loss.

I call the pump man if there's a problem with the pump. It's set to a depth of 180 feet, and it requires specialized machinery, which I don't have, to pull it out of the ground. The other specialist I'll call in is the refrigeration man if I'm having a problem with my walk-in cooler. Other than that, I do everything myself. Many of the skills that I need I learned from my father, others by trial and error. Increasingly, YouTube is another helpful source.

The week goes like this: the trailer has a flat tire; the blackberry trellis got lodged in the wind and needs to be straightened back up; there's a pool of hydraulic fluid on the ground where the tractor was parked; there's a swampy area in a far field where creatures have damaged a pipeline; a bearing has failed on the harrow, and as long as you're pulling it to pieces you might as well replace all eight bearings; the shut-off valve for the water tank is leaking; and the compressor keeps tripping a circuit breaker. And so you fix all these things on your list, and the next week you have a new list, and the week after that, another list. It doesn't end.

Keep in mind that one doesn't have to be a master—just a jack. The work doesn't have to be perfect—just adequate. And as long as it's adequate, there's a certain satisfaction to be had in competently fixing everything that breaks. But it does require some skills, and some courage.

Planning

The students at the California Farm Academy are required to present a business plan for their proposed farming operations. This includes a mission statement, an elaboration of goals, a synopsis of market research, and a projection of income and expenses for each of the first five years. Not only is such a plan useful for forcing the farmer to think through his project, but it is also a requirement when applying for an agricultural loan.

Preparing this type of business plan is standard procedure in any business school in North America. The student writes out an elaborate hierarchy of missions, goals, strategies and tactics, and tapes it to his bathroom mirror so that he can review it every morning while he brushes his teeth. The concept of setting and relentlessly pursuing goals is so pervasive in our culture, not only in business but also in personal life, that to question it seems an eccentricity.

I am not an enthusiast for this goal-setting approach. It strikes me as having hidden psychological and philosophical costs, and carried to excess it makes for an unattractive person. I have never had a business plan for my farm, nor have I ever applied for an agricultural loan, since I operate on a cash basis. Neither do I have a mission statement or goals. I have an approximate idea of what I'm doing, and I'm always paying attention. It's somewhat like being in a small boat in a strong current. The goal-setter picks up the oars and starts rowing vigorously in the direction of his predetermined goal, but I would rather concede the greater power of the current, and study it, and make an adjustment now and again with the oars so that I can reach a suitable landing, even if it wasn't the one I originally intended.

People have asked me how I have come to make various decisions about my farm and chosen a course of action. A truthful answer would require me to admit that I am afflicted by moodiness. In manic phases I undertake ambitious projects, unpremeditated and poorly conceived, sometimes making irreversible commitments. Then in a more steady state of mind I deal with whatever I had got myself into. And in darker times I study how I might escape obligations I have unwisely brought upon myself. As I have gotten older, the amplitude of these changes in mood is less than it used to be, and I'm more consistent. Nonetheless, it strikes me that the unwavering pursuit of goals requires a much more steady personality than I have. This is an area in which a corporate farm will have a more even course. It is unlikely that all of the members of the board of directors will be manic at the same time, and so a fairly sober and considered plan can be followed.

Along these lines, it is worth noting that a small family farm will follow an arc of productivity over its lifespan. In the early years the farmer is inexperienced, and makes mistakes. By mid-career he is master of his occupation. And later in life the farm goes into decline as the farmer's strength diminishes, his ambition wanes, and his cognitive skills become increasingly sketchy.

CHAPTER 14

Putting the Pieces Together

Conventional commodity agriculture in North America uses the most extreme technologies available: driverless tractors guided by GPS; variable-rate applicators of fertilizers and pesticides that integrate GPS and GIS to deliver varying amounts of product foot by foot over the farm as needed; irrigation systems that use buried moisture sensors transmitting data to a computer that in turn regulates pumps and solenoids to provide precision irrigation; and robotic devices with optical sensors that can knock out weeds from the crop while the tractor traverses the field at a good rate of speed. There are several forces that push the commodity farmer in this direction. His advisors—personnel from the departments of agriculture and the land grant universities—advocate this approach. Insurers and lenders who finance the operation may demand it. The buyers of commodities require a uniform and predictable product as an input for manufactured foods. And finally, in the context of long-entrenched policies of cheap energy/expensive labor, this is often the most profitable way to farm over the short term.

This extreme-tech approach to farming is sensitive to scale, and has developed in a way that makes it suitable only for large enterprises. The high capital costs must be spread over many acres to be justified. Moreover, technological complexity, at its current stage of development, must be harnessed together with biological simplicity. It is designed to work with a monoculture of a single genotype under the philosophy of "kill everything except the crop."

Appropriate Technology

On a small farm such as mine, extreme technologies are not suitable purely as a matter of scale. But aside from that, the philosophy of embracing biological complexity as a means to a robust and resilient ecosystem also dictates a simple and flexible technology as most appropriate, including considerable hand labor. For example, in determining when and how much to irrigate, I do not make use of a complex computerized analysis. Rather, I look at condensation on the leaves at dawn, and feel the soil, and look at the color and angle of the leaves in the heat of the afternoon, and observe the lengths of shoots, and combine these with other subtle judgments to make my decision. It is an artisanal rather than industrial approach, and as with other artisanal occupations it requires a long apprenticeship to achieve competence.

Consider, for a moment, a purely artisanal society, perhaps in a prosperous little valley in the mountains of Peru. The occupations are farmer, herder, weaver, tailor, miller, baker, potter, woodworker, herbalist, and a few others. All of these are skilled occupations, and require thousands of hours of practice to achieve mastery. No one person masters more than one, or rarely two, trades, and so the society is integrated by mutual dependency on one another's skills. An artisanal method is not merely a matter of a type of production—it is also a mechanism of social cohesion. And this applies to farming as well; a farmer supplying her local market, respected for her competence and probity and known to her customers, strengthens her community by her labors.

Industrial production values capital over labor, and favors uniformity, anonymity, and predictability over individuality. And if we ask, "Who benefits?" the answer clearly is capital, not labor. As industrial agriculture continues to develop, the role of the farmer is steadily diminishing, as is the demand for his skills. He is transformed from an artisan to an industrial worker, and he may end up being merely the attendant of a collection of devices.

Industrial production has its place. I would not want an artisanally produced tractor—idiosyncratic and unique, and requiring custom-built parts for repairs. I want an industrially produced tractor that incorporates the principal virtues of industrial products: uniformity and predictability. But for simpler tools, I prefer one made by the local blacksmith over one coming from an overseas factory.

To return to the question of appropriate technology, the idea that a biologically diverse small farm is best managed in an artisanal style implies simple technologies and an abundance of hand labor. The creation of jobs on farms would seem to be socially and economically desirable. However, the policies of the state that demand high wages and high overhead dictate that if the artisanal farmer is to hire labor, then he must sell his products at a high price, and in doing this he is forced into supplying the wealthier customers and ignoring the poorer ones. That is, the benefits of community cohesion and solidarity that derive from artisanal occupations are not extended to the whole community, but only to its wealthier portion. In addition to discouraging labor-intensive undertakings, the state's policies specifically favor industrial production; these include deliberate underpricing of energy, tax policies generous to industry and capital, outright subsidies of commodity crops, ethically dubious machinations of finance, uncompensated environmental degradation of many kinds, and the privatization of profits combined with the socialization of losses.

From a purely economic point of view, the most appropriate technology is almost always the most extreme and advanced technology that one can afford. If one's criteria are not purely economic, but include ecological health of the farm, then it is often the case that a simpler technology is more appropriate. And beyond economics and ecology there is a societal criterion that may dictate the simplest technologies of all. Because of automation, opportunities for simple, meaningful work are being destroyed at an accelerating pace, leaving many without work. Europe is ahead of us in this regard, but our direction is clear. The most palatable political solution is some form of guaranteed income—a welfare state. But welfare is destructive of the recipient's sense of self-worth, and of family life; the fatherless family is an emblem of the welfare state. It would be better to use some of those resources to support labor-intensive, low-tech farms on urban fringes, offering meaningful work to otherwise marginalized citizens. From this perspective, the very simplest technologies may be most appropriate.

Sustainability

Putting the operation of the farm into a larger spatial and temporal context requires some consideration of sustainability. The widespread preoccupation with this term is a measure of the extent to which people are coming to recognize that the condition of the planet is increasingly precarious. The search for what is sustainable is seen as the route to avoiding or delaying catastrophic change.

Like the words "ecological" and "environmental" in the 1970s, or "organic" in the 1990s, the word "sustainable" has been used in so many different ways and made to serve so many conflicting agendas that its meaning has become very imprecise. ("Resilient" appears to be next in line for this kind of semantic abuse.) The general idea is that a sustainable practice is one that can be carried out indefinitely without leading to its own destruction. In this discussion, I will make use of the lucid account of sustainability that was put forth by Paul B. Thompson in *The Spirit of the Soil* (1995).

We hear people refer to an "organic apricot." This is a sloppy abbreviation of speech for what should be, "an apricot produced by organic farming practices." There is no feature of the apricot itself that makes it organic or not. Similarly, in the case of sustainable farming, what is sustainable (or not) is the farming system, that is, the integrated collection of practices by which the farm operates.

In evaluating a farming system in terms of sustainability, several variables must be defined in advance: the time period in question; the boundaries of the system (not just geographic boundaries but also the inclusion or exclusion of personnel, regulations, materials, markets, etc.); and the criteria by which sustainability is to be judged. How these variables are defined will very much influence the conclusions we reach about sustainability. In the case of my farm, I propose to make the analysis at three scales, using differing definitions of the boundary conditions and of the criteria by which the outcome is judged. In each case the time frame to be considered is forty years—the typical working lifespan of a small-scale farmer.

FIRST ANALYSIS: CROP ECOLOGY

This is the narrowest interpretation. The geographic boundary includes only the cultivated land. Crops, soils, and water are included in the farming system, but the farmer, farm economy, farm social context, and energy inputs are not considered. Compost and seeds (including cover crop seed) are outside the system. The criterion for sustainability is that the yield from the crops (adjusted for extreme weather) does not decrease over the time interval considered.

In this analysis, the farm operation is clearly sustainable. There is no soil erosion. There are no applied poisons that might accumulate over time. Yields continue to rise for the perennial crops, and are consistent for the annual crops. Soil fertility is static or increasing. In principle, this farming system could be continued indefinitely into the future without problem.

SECOND ANALYSIS: THE FARM AS AN ECOLOGICAL, ECONOMIC, AND SOCIAL ENTERPRISE

The geographic boundary is the legal boundary of the farm, and includes the cultivated lands, buildings, roads, and riparian forest. Inputs of energy, water, and materials are included in the farming system, as are the markets for farm products, the regulatory environment, and the social context of the farm. I use four criteria for sustainability: a steady or increasing cumulative annual soil respiration as a measure of ecosystem health, an increase in biodiversity, a decreasing use of fossil fuels, and a positive economic and psychological well-being of the farm family.

I do not have a direct measure of soil respiration. I infer it from the abundance and activity of soil organisms and from a complex three-dimensional soil structure. Soil compactions, extremes of soil chemistry, and accumulation of toxic substances, including salts, will suppress soil respiration; high organic matter and good soil aeration will enhance it. By these surrogate measures, soil respiration is increasing year by year, especially in the land devoted to perennial crops where the soils have high organic matter and excellent structure. The abundance of mushrooms after a rain, the ubiquity of earthworms, and the complex architecture of the soil are indicators of robust soil health. In the cultivated land devoted to annual crops, these indicators are less robust. Before I owned this land it had been farmed by destructive industrial methods, including excessive cultivation, frequent use of poisons, and injections of anhydrous ammonia as a fertilizer. I did not measure soil organic matter at the outset, but I believe that it is slowly increasing in the cultivated land, although still low. Tillage may cause a temporary spike in soil respiration as organic matter is rapidly oxidized, but in the longer time frame it decreases soil respiration by destruction of soil structure and soil organisms. My experiments with strip tillage and no-till for annual crops have so far been unsuccessful, but I haven't given up yet. In summary, by the criterion of soil respiration, the farming system can be considered to be sustainable.

Compared to what I started with—an open field with a long history of destructive industrial farming of monoculture crops—biodiversity is tremendously increased, both in the number of species present and in the size of populations. It took more than a decade for frogs and toads, which are sensitive indicators of ecological health, to recolonize the farm, where they are now common. I was advised that many of the bird species in the region are limited by a shortage of nesting sites, and so, together with neighbors, we have installed numerous nesting boxes of various kinds, nearly all of which are regularly occupied. The insect fauna has become so diverse and interesting that it distracts me from my chores. Many of the wild species on the farm require suitable habitat on a much larger scale than our farm in order to prosper, but at least for the land under our care, the high level of biodiversity indicates that our practices are sustainable.

The energy budget of the farm has been described in chapter 9. Fossil fuels are required to operate the farm, and while their productivity is greater than on many conventional farms, to be less bad than the alternatives is a meager sort of virtue. The transition away from fossil fuels cannot happen rapidly. Suppose, for example, that I replace my car and van with electric vehicles, together with an array of photovoltaic panels to charge their batteries. On a day-to-day basis, gasoline is no longer required. However, the manufacture of those vehicles and photovoltaic panels, including working their materials—steel, silicon, glass, plastic, rubber—as well as their assembly and distribution, is based on fossil fuels, which must be considered a fossil fuel cost of their operation. It is probably less costly of fossil fuels for me to continue using my old vehicles, which are likely to last for my remaining lifetime, than to switch to electric ones. As long as one uses machinery, the pace at which one can fully abandon fossil fuels is determined largely by the pace at which the entire material economy can give up fossil fuels.

The on-farm requirements for propane, gasoline, and diesel fuel could be supplied by using less than 10 percent of the olive oil produced as a biofuel, although to use extra virgin olive oil for biodiesel would be an appalling crime. Alternatively, some open ground could be planted to dry-farmed safflower, which in this area yields about 200 gallons of oil per acre. Daily use of fossil fuels could be eliminated, but the fossil fuel costs of the embodied energy of farm equipment is beyond our control.

On a year-to-year basis, the economics of the farm are sufficient to support the family without off-farm income. Although the wages are low by American middle-class standards, the income is enough to supply life's necessities and a few of the luxuries. By global standards, the income would be considered generous. One of the reasons that the income works for us is that we have no ambition to become wealthy; our notion of "enough" is a modest one.

However, looking at a longer time frame, the economics of the farm appear to be less propitious. We started out with a strong work ethic and ten cents in our pockets, and no likelihood of financial support from our families. We both worked at interesting but low-paying jobs, usually with a second job on the weekends. This continued for years, and by the time we were able to afford to buy the farm, I was 39 years old and Dianne was 34. We had acquired many useful skills along the way, but nonetheless this is rather late in life to be starting farming.

The increasing price of farmland in this area has far outstripped the rising prices of other assets, most notably labor, to the extent that a young couple starting out today has no possibility of replicating our experience. Their cumulative lifetime income would not suffice to buy twenty acres. How did this come to pass? Partly it is a simple question of supply and demand; the supply of farmland is decreasing as fields are lost to urbanization, while the demand is increasing. It is also true that the price of farmland reflects not only its agricultural value, but also its value as an instrument of financial speculation and a place for a homesite; it is the latter two that primarily drive the price. Farmland investment hedge funds with sacks of money are buying up farmland in the valley without much concern about the price. Most of this land is subsequently farmed by corporate tenants. In my neighborhood, only two of the eleven parcels are owned by farmers. The rest are the properties of two-income households of professionals who wanted to live in the country. It's unlikely that these professionals considered that by bidding up the price of farmland they were denying beginning farmers a livelihood.

In this context, our farming system is not sustainable. When we retire, there is no possibility that someone of the next generation could repeat our experience on this land. A young farmer could start up on an appropriate piece of land with a long-term lease, and this is most likely the way of the future. Using leased land has certain advantages. Capital that would have been used to purchase land is freed for purchasing equipment and building farm infrastructure. And if the farmer lives in town and commutes to his farm, his family is spared the social isolation and oversized carbon footprint entailed in living in the country. However, much of the increased wealth that has accrued from our labors did not show up as income, but rather as increased value of the farm. The operator farming on leased land is denied this opportunity. And depending on his relationship to his landlord, he may be unable to justify investment in perennial crops or in farm buildings (greenhouse, barn, cool storage, packing shed) that will not legally be his, and consequently he may have to struggle along with inadequate facilities. And the farmer may be discouraged from putting funds and labor into soil-building activities, the benefits of which accrue in an indistinct future, if his lease is short-term or insecure.

A minor but related point is that much of our building construction (furniture too, for that matter) used recycled materials salvaged from abandoned buildings. When I first started salvaging old buildings it was considered an eccentricity; now it is a recognized and competitive business with many players, and the few suitable buildings available are sold at high prices. Salvaging, for me, came at a propitious time, which has now passed, and to that extent our experience could not be replicated.

Another aspect of economic sustainability is the question of whether the farm has generated enough financial momentum to allow us to retire at some point when we no longer have the strength or the will to keep farming. We could lease out the land and farm buildings, excluding the house, for enough to cover the cost of property taxes and maintenance of the farm infrastructure. We have enough savings to provide us a survival-level income, which we more or less expect to be augmented by Social Security income. That said, we will postpone retirement as long as we continue to enjoy what we're doing, and I expect to still be farming at age 80. Of course, we could sell the farm and cash out the accumulated wealth accrued from our decades of work, but we have no plan to do that. In an increasingly unstable world it is important to keep the farm as a refuge for family and friends in times of economic collapse and social disarray.

As a social enterprise, the farm functions well and should be able to continue to do so. The institutions in which the farm operates (farmers markets, stores, credit unions, schools, police and fire, agriculture department) are for the most part well designed and competently staffed. Democracy works best at the local level, and we expect that it will continue to do so. The social fabric into which the farm is woven is strengthened by our direct relationships with our customers at the market and by the many visitors to the farm, from school children to senior gleaner groups, whose visits leave them with a sense of solidarity with our enterprise.

THIRD ANALYSIS: REGIONAL SUSTAINABILITY

The region to be considered is roughly a circle of about thirty miles in diameter, reaching but not crossing the Sacramento River. The principal municipalities are Fairfield, Vacaville, Davis, Woodland, and West Sacramento. Most of the land is level agricultural land of the Sacramento Valley floor, but a portion of the coastal mountains is included. The criteria for sustainability include static or increasing agricultural yields, stable or increasing biodiversity, a balanced energy budget, and static or increasing economic wealth. As with the other analyses, the time frame is 40 years.

Unlike most agricultural regions, the soils here are not troubled by erosion because of the flatness of the land. The principal drainage of the area, Putah Creek, is a clear-water stream (after a heavy rain it may carry some soil washed out of the coastal mountains); in fact, the creek supports an annual migration and spawning of salmon. In the eastern half of the region, where rice is grown, flood irrigation with muddy waters of the Sacramento River may result in soil deposition—the opposite of erosion. Many of the soils planted to row crops are nearly lifeless as a result of decades of excessive tillage, overuse of pesticides, and application of caustic chemical fertilizers, but soil is resilient, and if treated well it can return to life within a couple of years.

The trend of regional agriculture is to replace row crops and field crops with orchards, principally nut crops. The orchards are ecologically more benign, with reduced tillage, reduced energy inputs, increasing sequestered carbon, and provision of habitat for wildlife. However, they also have increased demand for water. The classic row crop rotation in the region was tomatoes (irrigated) alternating with wheat (unirrigated). In a severe drought, all of the land could be idled. But with orchards, there is no unirrigated land, nor can irrigation be reduced in a drought without harm to the trees. The result is increased demand for water and decreased flexibility. Because delivery of surface water from irrigation districts is not guaranteed, the orchards are irrigated by pumping groundwater, and this has led to increasing overdrafting of the aquifer. The mid-summer static water level in my wells has dropped from 45 feet to 62 feet over a few decades, and in other areas the drop is more severe. Chronic overdrafting of the aquifer leads to collapse of the pores in the soil in which the water had been held, which precludes recovery of the aquifer, and which is manifested at the surface of the land by soil subsidence. This is clearly an unsustainable situation posing a long-term threat to the agricultural productivity of the region.

On a field-by-field basis, yields of crops have been steady or increasing. Intensification of farm practices, such as a shift from direct seeding to transplanting, a switch from furrow irrigation to drip irrigation, and the use of GPS precision guidance of machines, are contributing factors. Regionally, the cumulative increase is less than expected because of loss of farmland to urbanization. New statutes protecting farmland have been put in place, and the depredations of the 1980s and 1990s are unlikely to be repeated, although incremental loss continues. Visitors from abroad continue to be astonished by the way in which the world's best farmland has been covered over with low-density sprawl of shabby houses. Biodiversity of the region, while perhaps less than it was in aboriginal times (elk and antelope are no longer here), is stable or increasing. A number of natural habitats have been formally protected, and many are carefully managed and curated.

The energy budget of the region is highly dependent on imported energy. The principal crops—tomatoes and walnuts—have an energy efficiency ratio of less than one, although this does not include the accumulating biomass of the orchards. As orchards reach the end of their productive cycle (30 to 40 years), they are pulled out. The trees are chipped and the chips sent to a local energy plant that burns agricultural waste to generate electricity; this contributes only a tiny portion of the electrical demand of the region. There is a very minor amount of photovoltaic, hydroelectric, and wind generation of electricity, and a small production of natural gas. The combined contribution of these is just a few percent of the electricity use of the region. Fossil fuels for transport are imported to the region from distant sources. Overall, the energy captured by photosynthesis is less than the energy brought to the region in the forms of imported electricity and fossil fuels. Given the likely increasing prices and decreasing availability of imported energies due to declining reserves and global geopolitical instabilities, the energy base of the region must be considered to be unsustainable.

The principal industries of the region are agriculture and food processing; building construction; creation and sale of intellectual property, particularly in the area of biotechnology; light manufacturing; education; and government. Notable institutions are an Air Force base, a major land-grant university, and a prison. Many of the residents work in government offices in Sacramento, which lies just outside the region, augmenting the role of government in the region's economy. The emphasis on regulatory, bureaucratic, and military occupations is notable. The economy as a whole is dependent on funds derived from taxes and fees levied on creative enterprises outside of the region. While this is not necessarily unsustainable, it is much less robust than an economy based primarily on the production of useful objects and services.

In summary, at the regional level the drawdown of the underlying aquifer, the extreme dependence on imported energy, and the predominance of distributive rather than creative occupations, all suggest that in the long run the region's ecology and economy are not sustainable. Whether this is manifested in a 40-year time span or at some longer interval remains to be seen.

I have not thus far brought up the question of climate change, which is predicted to lead to increased statistical variance in weather in the context of diminishing total rainfall and increasing temperatures. This is clearly a threat to the agricultural basis of the regional economy, as well as to biodiversity. Returning for a moment to the second level of sustainability analysis, my farm is in better shape than many because the principal crop, olives, is very drought-tolerant, and is productive in regions far drier and hotter than this one.

Regional Farming Systems

In this work I have attempted to describe a farming system, that is, a collection of farming practices that are philosophically, ecologically and economically integrated. My farming system is atypical in that the principal aim is not to maximize profit, or yield, but to maximize quality of life. The modern obsession with yield and profit came to the fore after World War II; previous to that, especially going back to the 18th century, farmers were likely to be more interested in the challenges and satisfactions of self-sufficiency and in community solidarity. The way that I manage my farm results in a yield that is quite a bit less than the land is capable of if it were farmed with maximum intensity. Compensations for decreased yield are to be found in greater ecological health of the farm and increased biodiversity.

There are a few other farmers in my area who farm with philosophies and practices similar to mine, but collectively we control only a very tiny portion of the total farmed acreage. Most of the land is farmed according to the tenets of conventional agriculture, that is, agricultural practices recommended by the departments of agriculture, the land grant universities, Monsanto, Cargill, and other forces of the agro-industrial complex. These practices favor simplicity over complexity, chemistry over biology, profit over sustainability, and capital over labor. The goals are maximum yield and maximum profit, embedded in a reductionist notion of plant nutrition and in metaphors of warfare for the relationships to non-crop species. Much of the conventional agriculture is carried out by farming corporations that manage thousands or tens of thousands of acres, most or all of it leased land.

Between these extremes are a number of farms, mostly orchards, mostly operated by individuals, and mostly in the range of 50 to 500 acres that are operated with a sensitivity to ecology and sustainability. Although they are monocultures with a goal of maximum yield/profit, the farmers have planted hedgerows and wildlife corridors and have a light hand with pesticides. Legume green manure crops often are grown in preference to applying synthetic fertilizers, and a few tons of compost per acre may be applied each year. These farmers have an understanding of and concern for the health of the soil. This enlightened style of farming was started by a few forward-thinking individuals, and then imitated by many others.

It is not accidental that the classes of farming systems correspond to differences in scale. A commodity tomato grower can sell tomatoes to the cannery, profitably, at less than four cents per pound. He achieves this is part by using highly sophisticated machinery. A tomato harvester costs about $400,000; in order to justify that expense it should be running 24 hours a day from the start of harvest season to the end, which means the farmer needs at least 1,000 acres of tomatoes. And for that size of operation he needs big tractors ($300,000 each) and big implements to go with them. This technology is not scalable down to the 20-acre farm, and the small-scale farmer certainly can't compete on price with the commodity grower. And so the small-scale grower concentrates on fresh-market heirloom varieties, and maintains a highly diverse crop lineup in order to maximize income in his sales channels. In a sense, small scale dictates biodiversity while big scale requires monocultures. A summary of the farming systems is presented in the sidebar on page 148. While these are gross generalizations with a great many exceptions, they nonetheless are useful as a shorthand summary of the regional farming systems.

Optimizing Valley Farming

We can question what might be the optimal array of systems for the region as a whole. I believe that it would not work to convert the entire region to small farms, leading to an agrarian landscape such as that of Greece or Japan. For starters, the reconfiguration of land tenure would present monumental obstacles. And where would all these small-scale farmers come from? It would take a generation to recruit and establish them. The tremendous investment in commodity infrastructure and machinery could not simply be abandoned, and it has decades of useful life left. And what

Regional Farming Systems

BIG FARMS

1,000 acres plus
Leased land
Corporate tenant
Capital-intensive
Extreme technology
High inputs
Monoculture commodity crops
Conventional farming practices
Low per-unit price for products
Very low biodiversity
Decreasing soil carbon
Principal crop pests: Insects
High pesticide use
Sterile soil
Product fungible
Farmer anonymous
Market global
Political orientation: Conservative
Social orientation: Hierarchical
Obligations: Profits for shareholders
Goal: Short-term profits

MID-SIZE FARMS

50–500 acres
Owner/operator
Mildly capital-intensive
Mixed technologies
Medium inputs
Mostly orchard crops, monoculture
Enlightened practices (hedgerows, wildlife corridors, cover crops)
Medium per-unit price for products
Medium biodiversity
Static soil carbon
Healthy soil
Product fungible
Farmer anonymous
Market global
Obligations: Welfare of family
Goal: Long-term profits, sustainability

SMALL FARMS

5–30 acres
Owner/operator
Labor-intensive
Simple technology

Low inputs, low energy use
Diversified crop mix
Ecologically sound farming system
High per-unit price for products
High level of biodiversity
Increasing soil carbon
Principal crop pests: Vertebrates
No pesticides
Healthy soil

Product unique to farm
Farmer known to customers
Market local
Political orientation: Liberal
Social orientation: Egalitarian
Obligations: Healthy community, healthy ecosystem
Goal: High quality of life; thriving ecosystem

happens to the retail price of food when the farmgate value of processing tomatoes jumps from four cents per pound to a dollar? Expect great unhappiness and resistance from the end consumers.

What strikes me as ideal in the short run (ten years or so) is to have a scattering of small farms throughout the valley—perhaps 5 percent of the total area, creating an archipelago of islands of biodiversity, connected by wildlife corridors. The remaining land should be in middle-sized farms, with the question of amortizing the cost of expensive machinery to be addressed by shifting those functions (transplanting, harvesting) to contractors who would work on a number of farms. Much of the harvest of hay and small grains in the rest of the country is done in this way, as is some of the nut harvest, pruning, and brush chipping locally. While such a shift would represent an improvement in the ecological and economic health of the valley, there would be many obstacles. For example, increasingly farmland is owned by farmland investment corporations whose shareholders are wealthy urban people or institutions; they have no interest in sustainable practices, but rather demand maximum short-term profits.

Although a goal of 5 percent of the area in small farms seems very modest, I believe it would be difficult to achieve. Most of the small family farms like mine in this region have one of two fates: either they get big, scaling up to hundreds of acres and dozens of employees, or they fail. Of the numerous small farms launched in my area over the last thirty years, only three that I know of have persisted past year five. For the 5 percent goal to be reached, new institutions and new agrarian philosophies are called for. One could imagine small parcels of farmland, mostly at urban fringes, bought by churches, school boards, hospitals, food banks, or municipalities and farmed by salaried farmers with large crews of laborers to produce foods that such institutions currently purchase through traditional wholesale markets on behalf of their constituents. This would be both shrewd and righteous for those institutions.

Another approach is a vertically integrated cooperative venture, combining a small farm, a restaurant, a bakery, a catering service, perhaps a daycare center, and a venue for music and events. An organization like this has the advantage of a fluid labor pool that can be moved about to wherever labor is needed. There are several successful examples of this sort of venture around the country.

Or, imagine a more extreme possibility. All citizens reaching age 18 are required to perform two years of national service. This could be military service, urban renewal, care of wild lands, tutoring of needy children, or small-scale, labor-intensive farming of fruits and vegetables. The numerous small farms, scattered around the country, would be set on land leased or owned by the federal government, and would be managed by salaried head farmers. The foods produced would be distributed through a variety of channels—some traditional markets, some subsidized distributions—to the citizens, including those who are indigent. Of the young people putting in their time on the farms, a very few might be inspired to become farmers, and the rest would at least gain some appreciation of our relationship to the land. Such a program, even on a vast scale, could be operated at a cost of less than one percent of what we now spend on the military. Improved health of the population, resulting from the distribution of healthy foods produced on the small farms, might in itself compensate the costs. Of course, there would be a lot of resistance to the creation of such a program, but the first step to creating a worthy institution is to imagine it.

Future of the Farm

We consider it a great privilege to have been entrusted with the care of a beautiful and productive parcel of land all these years, but we can't keep it up forever, and at some point we'll be ready to retire. Our daughters love the farm as their lifelong home, but they are not interested in operating it as a business. For a variety of reasons we do not intend to sell the farm, and so eventually we will need to find someone to take it over on a lease. We have made friends with a young couple who are intelligent, ethical, and hard-working, and who share our values, and we are attempting a gradual transfer of the farm operation to them. For a few years they had a garden on the farm. Then they leased 150 olive trees from us, and began to develop their skills at olive culture, at operating the milling machinery, and at marketing.

The following year they leased 300 trees from us, and established their own brand of olive oil. The idea is that we will gradually slow down over a period of years while they gradually ramp up (at this point they still have day jobs). Ultimately it will be their farm, although we will hold title to the land. Outside of the olives, it's likely that they will develop their own strategies of farming that will no doubt be different from ours. This is not a traditional mechanism of farm succession, but it is appropriate to the times. We wish them success.

Epilogue

In this work I have described the operation of a small farm over a period of three decades. The solutions, or partial solutions (or in a few cases, continuing failures) that I have come up with for various agronomic and horticultural problems may be relevant and useful to others. But in the broader sense of the farm's agroecology, this account must be considered more of a historical curiosity than a guide for starting farmers. Circumstances—technology, economics, land tenure, regulation, markets, social institutions—have changed so much that our history cannot be recapitulated by others, and practices that have worked well for us may no longer be adaptive.

The beginning farmer faces new challenges. Access to land is increasingly difficult and expensive. Traditional access to markets is also difficult: the best farmers markets may have waiting periods of a decade or more before a spot becomes available. At the same time, the ubiquitous burden of debt from student loans saps the new farmer's income stream and impairs her access to capital. On the other hand, the internet may facilitate entry to new markets, and new forms of cooperative enterprise create opportunities. The rules of the game are always changing, but the determined individual will probably be able to find a path to a satisfying life of small-scale farming.

In some of the post-industrial states (California, the Netherlands, Japan) the obstacles for a starting farmer are so great that emigration is an appropriate alternative. For me, the promised land always was Chile—far enough south that the climate is more oceanic than continental, with abundant water, inexpensive land, and sparse population, and at the same time remote from the expanding social turmoil of the northern hemisphere. But such a path is suited to someone in their twenties, not their seventies. At this stage, it is not for me, but I would encourage a young person who is stymied at home to pursue that adventure.

An Abbreviated Almanac

Some activities continue throughout the year: maintaining equipment, filing paperwork, weeding, pruning, irrigation (even in rainy weather seedlings in the greenhouse may need irrigation), cleaning, and hauling trash. Other activities, which span a briefer interval, are mentioned in the almanac, although the list is far from complete.

JANUARY

Finish harvesting olives for oil
Dismantle and clean olive mill (two days)
Harvest olives for Kalamata-style table olives, salt-cured table olives
Service tractors
Lubricate implements
Prune persimmon trees
Prune olive trees
Harvest flowering branches, *Prunus mume*
Harvest anemones
Harvest citrus fruit (oranges, bergamots, lemons, kumquats)
Make citrus marmalades
Make soap, hand salve
Weed row crops
Dormant oil spray to nectarines
Order seeds
File crop reports

Renew organic certification
File sales tax report

FEBRUARY

Prune olive trees
Harvest citrus (oranges, bergamots, lemons, kumquats)
Make citrus marmalades
Make soap, hand salve
Plant sunflowers (greenhouse)
Plant scabiosa (greenhouse)
Plant lettuce, broccoli, radishes, beets
Harvest flowering quince
Harvest anemones
Harvest early gerbera
Harvest lettuce
Weed row crops
File income tax
Submit nitrogen management plan

MARCH

Finish pruning olive trees
Haul in firewood
Split firewood
Schedule chipping contractor for olive grove
Harvest anemones
Harvest ranunculus
Harvest tulips
Harvest iris
Harvest peonies
Harvest gerbera
Harvest oranges
Work up beds for spring planting

Spread compost
Spread gypsum
Transplant tuberose
Plant sunflowers (weekly)
Harvest citrus (oranges, kumquats)
Harvest lettuce
Harvest green onions
Start seeds for summer annuals (greenhouse)
Set squirrel traps
Set new irrigation for annual crops
Repot gerbera
Fertilize gerbera
Start seeds of cucumber, melons, tomatoes, onions, peppers (greenhouse)
Tie up blackberry canes

APRIL

Mow cover crops
Work up planting beds
Plant sunflowers
Start melons (greenhouse)
Plant cucumbers
Transplant tuberose
Transplant summer annuals
Repot gerbera
Harvest peonies
Harvest ranunculus
Harvest anemones
Harvest alstroemeria
Mow remaining cover crops
Work up planting ground
Plant summer annuals (greenhouse)
Transplant summer annuals
Charge and inspect irrigation lines

Begin irrigation cycles
Order seeds
Weed peonies
Weed annual crops
Spread mulch

MAY

Harvest peonies
Harvest alstroemeria
Harvest sunflowers
Harvest annual flowers
Harvest grapefruit
Harvest blackberries
Make blackberry jam
Plant sunflowers
Plant melons
Plant cucumbers
Plant beans
Transplant annual flowers
Transplant peppers, tomatoes, eggplant, onions
Weed annuals
Weed peonies
Mow olive orchard
Harrow olive orchard
Irrigate orchards, row crops
Trap squirrels, gophers

JUNE

Plant sunflowers
Plant melons
Plant cucumbers
Start annual flowers (greenhouse)

Transplant annual flowers
Harvest sunflowers
Harvest cucumbers
Harvest other flowers
Harvest lavender
Harvest blackberries
Make blackberry jam
Harvest apricots
Make apricot jam
Weed row crops
Irrigate, irrigate, irrigate
Monitor olive fruit fly
Bait for fruit fly if needed
Spray olives with kaolin clay, if needed
Order tulips, iris, anemone bulbs

JULY

Cut out old blackberry canes
Tie up new blackberry canes
Prune apricots
Prune other stone fruits
Plant cucumbers
Plant watermelons
Plant cantaloupe
Plant sunflowers
Plant annual flowers
Harvest melons
Harvest cucumbers
Harvest early apples
Harvest annual flowers
Harvest tuberose
Weed row crops
Irrigate orchards

Irrigate row crops
Mow apricots
Trap squirrels, gophers

AUGUST

Harvest sunflowers
Harvest other flowers
Harvest melons
Harvest cucumbers
Harvest apples
Harvest figs
Make fig jam
Dry figs
Harvest plums
Harvest damsons
Make plum jam
Plant sunflowers
Plant annual flowers
Cut back lavender
Tie up blackberry canes
Weed row crops
Irrigate
Trap squirrels, gophers
Collect seeds, melons
Collect seeds, annual flowers
Send water, soil samples to lab
Pay estimated tax
Make soap

SEPTEMBER

Renew processor licenses
Harvest sunflowers

Harvest other flowers
Harvest melons
Harvest cucumbers
Harvest apples
Start overwinter flowers
Plant garlic
Plant onions
Pull up drip tape
Mow crop residues
Spread gypsum
Till soil
Spread compost
Prepare beds for fall/winter crops
Get cover crop seed
Plant cover crops
Harvest figs
Make fig jam
Dry figs
Harvest damsons
Make damson jam
Dry damsons
Irrigate

OCTOBER

Harvest sunflowers
Harvest other flowers
Plant over-wintering flowers
Plant spinach
Plant lettuce
Mow crop residues
Cultivate
Plant cover crops
Remove drip lines

Harvest apples
Begin olive harvest
Make olive oil (daily)
Harvest green olives for curing
Trap squirrels, gophers

NOVEMBER

Harvest olives
Extract olive oil
Harvest gerbera, tuberose
Harvest persimmons
Harvest pomegranates
Harvest mandarins
Harvest apples ('Pink Lady')

DECEMBER

Plant tulips
Plant iris
Plant anemones
Plant ranunculus
Harvest persimmons
Harvest apples
Harvest olives
Make olive oil (daily)

Background Literature

Altieri, M. *Agroecology: The Scientific Basis of Alternative Agriculture*. Boulder: Westview Press, 1987.

Barbour, M. and Major, J. *Terrestrial Vegetation of California*. New York: J. Wiley & Sons, 1977.

Bayliss-Smith, T.P. *The Ecology of Agricultural Systems*. Cambridge: Cambridge University Press, 1982.

Boserup, E. *The Conditions of Agricultural Growth*. Chicago: Aldine, 1965.

Bromfield, L. *Malabar Farm*. New York: Harper & Bros., 1947.

Brush, S.B. *Farmers' Bounty*. New Haven: Yale University Press, 2004.

Clark, C. and Haswell, M. *The Economics of Subsistence Agriculture*. London: Macmillan, 1964.

Clements, D. and Shrestha, A. (eds). *New Dimensions in Agroecology*. Binghamton, NY: Food Products Press, 2004.

Faulkner, E.H. *Plowman's Folly*. Norman: University of Oklahoma Press, 1943.

Freyfogle, E.T. (ed). *The New Agrarianism*. Washington: Island Press, 2001.

Fukuoka, M. *The Natural Way of Farming*. Tokyo: Japan Publications, 1985.

Gliessman, S.R. *Agroecology*. New York: Springer Verlag, 1990.

Hillel, D.J. *Out of the Earth*. New York: The Free Press, 1991.

Jackson, W. *New Roots for Agriculture*. Lincoln: University of Nebraska Press, 1980.

Jackson, W. *Nature as Measure*. Berkeley: Counterpoint, 2011.

Kirschenmann, F.L. *Cultivating an Ecological Conscience*. Berkeley: Counterpoint, 2010.

Kloppenburg, J.R. *First the Seed*. Second Ed. Madison: The University of Wisconsin Press, 2004.
Kowal, J.M. and Kassam, A.H. *Agricultural Ecology of Savanna*. Oxford: Clarendon Press, 1978.
Illich, I. *Energy and Equity*. New York: Harper and Row, 1974.
National Research Council. *Soil and Water Quality*. Washington: National Academy Press, 1993.
Netting, R.M. *Smallholders, Householders*. Stanford: Stanford University Press, 1993.
Pimental, D. and Pimental, M. *Food, Energy, and Society*. London: Edward Arnold, 1979.
Pope Francis. *Laudato Si, Encyclical Letter of May 2014*. The Vatican.
Prieto, P. and Hall, C.A.S. *Spain's Photovoltaic Revolution: The Energy Return on Investment*. New York: Springer Verlag, 2012.
Roeding, G. *Roeding's Fruit Growers' Guide*. Fresno, CA, 1919.
Sessions, G. (ed). *Deep Ecology for the 21st Century*. Boston: Shambala, 1995.
Shideler, J.H. (ed). *Agriculture in the Development of the Far West*. Washington: The Agricultural History Society, 1975.
Smaje, C. *Smallfarmfuture.org*
Smil, V. *Energy, Food, Environment*. Oxford: Clarendon Press, 1987.
Spaeth, H-J. W. "Dryland Wheat Farming on the Central Great Plains: Sedgwick County, Northeast Colorado," in *Turner and Brush*, 1987.
Tainter, J.A. *The Collapse of Complex Societies*. Cambridge: Cambridge University Press, 1988.
Thompson, P.B. *The Spirit of the Soil*. New York: Routledge, 1995.
Tivy, J. *Agricultural Ecology*. Harlow: Longman Scientific and Technical, 1990.
Turner, B.L. and Brush, S.B. *Comparative Farming Systems*. New York: The Guildford Press, 1987.
Vitek, W. and Jackson, W. *Rooted in the Land*. New Haven: Yale University Press, 1996.
Worster, D. *The Wealth of Nature*. New York: Oxford University Press, 1993.

About the Author

Dianne Madison

Mike Madison lives with his wife, Dianne, in Winters, California, where they operate a small, diverse organic farm, growing olives, apricots, citrus, melons, and a variety of flowers. He is the author of four books, including *Fruitful Labor*, *Blithe Tomato*, and *Walking the Flatlands*.